Building Planning and Drawing

AUTHOR

Dr. Aparna Dwivedi
Assistant Professor
Sir J.J. College of Architecture
Mumbai-400001, Maharashtra

REVIEWER

Prof. Pradeep Kumar
Professor,
Department of Civil Engineering,
Harcourt Butler Technical University
Kanpur-208002, Uttar Pradesh

KHANNA BOOK PUBLISHING CO. (P) Ltd.
4C/4344, Ansari Road, Darya Ganj,
New Delhi, India

All India Council for Technical Education
Nelson Mandela Marg, Vasant Kunj,
New Delhi, India

ISBN: 978-93-5538-437-9
Book Code: DIP468EN

Building Planning and Drawing
by Dr. Aparna Dwivedi

First Edition: 2025

Originally Published by:
All India Council for Technical Education, Ministry of Education, GOI

Exclusively Published & Distributed by:
Khanna Book Publishing Co. (P) Ltd.
4C/4344, Ansari Road, Daryaganj,
New Delhi (INDIA)

Visit us at: **www.khannabooks.com**
Write us at: **contact@khannabooks.com**

To view complete list of books, Please scan the QR Code: KPH

Printed in Bharat.

प्रो. टी. जी. सीताराम
अध्यक्ष
Prof. T. G. Sitharam
Chairman

सत्यमेव जयते

आज़ादी का अमृत महोत्सव

अखिल भारतीय तकनीकी शिक्षा परिषद्
(भारत सरकार का एक सांविधिक निकाय)
(शिक्षा मंत्रालय, भारत सरकार)
नेल्सन मंडेला मार्ग, वसंत कुंज, नई दिल्ली–110070
दूरभाष : 011–26131498
ई–मेल : chairman@aicte-india.org

ALL INDIA COUNCIL FOR TECHNICAL EDUCATION
(A STATUTORY BODY OF THE GOVT. OF INDIA)
(Ministry of Education, Govt. of India)
Nelson Mandela Marg, Vasant Kunj, New Delhi-110070
Phone : 011-26131498
E-mail : chairman@aicte-india.org

FOREWORD

Engineers are the backbone of any modern society. They are the ones responsible for the marvels as well as the improved quality of life across the world. Engineers have driven humanity towards greater heights in a more evolved and unprecedented manner.

The All India Council for Technical Education (AICTE), have spared no efforts towards the strengthening of the technical education in the country. AICTE is always committed towards promoting quality Technical Education to make India a modern developed nation emphasizing on the overall welfare of mankind.

An array of initiatives has been taken by AICTE in last decade which have been accelerated now by the National Education Policy (NEP) 2020. The implementation of NEP under the visionary leadership of Hon'ble Prime Minister of India envisages the provision for education in regional languages to all, thereby ensuring that every graduate becomes competent enough and is in a position to contribute towards the national growth and development through innovation & entrepreneurship.

One of the spheres where AICTE had been relentlessly working since past couple of years is providing high quality original technical contents at Under Graduate & Diploma level prepared and translated by eminent educators in various Indian languages to its aspirants. For students pursuing 2nd year of their Engineering education, AICTE has identified 87 books, which shall be translated into 12 Indian languages - Hindi, Tamil, Gujarati, Odia, Bengali, Kannada, Urdu, Punjabi, Telugu, Marathi, Assamese & Malayalam. In addition to the English medium, books in different Indian Languages are going to support the students to understand the concepts in their respective mother tongue.

On behalf of AICTE, I express sincere gratitude to all distinguished authors, reviewers and translators from the renowned institutions of high repute for their admirable contribution in a record span of time.

AICTE is confident that these outcomes based original contents shall help aspirants to master the subject with comprehension and greater ease.

T.G. Sitharam

(Prof. T. G. Sitharam)

ACKNOWLEDGEMENT

The author is grateful to the authorities of AICTE, particularly Prof. T. G. Sitharam, Chairman; Dr. Abhay Jere, Vice-Chairman; Prof. Rajive Kumar, Member-Secretary and Dr. Amit Kumar Srivastava, Director, Faculty Development Cell for their planning to publish the books on *Building planning and Drawing*. We sincerely acknowledge the valuable contributions of the reviewer of the book Prof. Pradeep Kumar for giving valuable inputs to complete the book.

I want to thank Ms. Pallabi Jana, Mr. Manas Karnjkar, Mr. Kunal Sawant and Mr.Atharv Jadhav for giving me the support and helped me in making drawings for the book. I also thank all my students and faculty members for these years whose feedbacks have proved valuable in putting the contents of this book.

This book is an outcome of various suggestions of AICTE members, experts and authors who shared their opinions and thoughts to further develop the engineering education in our country. Acknowledgements are due to the contributors and different workers in this field whose published books, review articles, papers, photographs, footnotes, references and other valuable information enriched us at the time of writing the book. Hope, this book will help the students understand the subject easily and conveniently.

Dr. Aparna Dwivedi

PREFACE

The book titled "Building Planning and Drawing" is an outcome of the rich experience of our teaching of basic architecture courses. The initiation of writing this book is to expose basics of drawings and the fundamentals of planning as well as enable them to get an insight of the subject. Keeping in mind the purpose of wide coverage as well as to provide essential supplementary information, we have included the topics recommended by AICTE, in a very systematic and orderly manner throughout the book. Efforts have been made to explain the fundamental concepts of the subject in the simplest possible way.

During the process of preparation of the manuscript, we have considered the various standard text books and accordingly we have developed sections like critical questions, solved and supplementary problems etc. While preparing the different sections emphasis has also been laid on definitions and laws and also on comprehensive synopsis of building codes for a quick revision of the basic principles. The book covers all types of medium and advanced level problems and these have been presented in a very logical and systematic manner. The gradations of those problems have been tested over many years of teaching to a wide variety of students.

Apart from illustrations and examples as required, we have enriched the book with numerous problems in every unit for proper understanding of the related topics. Under the common title "Planning" there are various topics covered in this book.

We sincerely hope that the book will inspire the students to learn and discuss the ideas behind basic Planning techniques and also to draw the building with various scales and views. This will surely contribute to the development of a solid foundation of the subject. We would be thankful to all beneficial comments and suggestions which will contribute to the improvement of the future editions of the book. It gives us immense pleasure to place this book in the hands of the teachers and students. It was indeed a big pleasure to work on different aspects covering in the book.

Dr. Aparna Dwivedi

OUTCOME BASED EDUCATION

For the implementation of an outcome based education the first requirement is to develop an outcome based curriculum and incorporate an outcome based assessment in the education system. By going through outcome based assessments, evaluators will be able to evaluate whether the students have achieved the outlined standard, specific and measurable outcomes. With the proper incorporation of outcome based education there will be a definite commitment to achieve a minimum standard for all learners without giving up at any level. At the end of the programme running with the aid of outcome based education, a student will be able to arrive at the following outcomes:

Programme Outcomes (POs) are statements that describe what students are expected to know and be able to do upon graduating from the program. These relate to the skills, knowledge, analytical ability attitude and behaviour that students acquire through the program. The POs essentially indicate what the students can do from subject-wise knowledge acquired by them during the program. As such, POs define the professional profile of an engineering diploma graduate.

National Board of Accreditation (NBA) has defined the following seven POs for an Engineering diploma graduate:

PO-1. Basic and Discipline specific knowledge: Apply knowledge of basic mathematics, science and engineering fundamentals and engineering specialization to solve the engineering problems.

PO-2. Problem analysis: Identify and analyses well-defined engineering problems using codified standard methods.

PO-3. Design/development of solutions: Design solutions for well-defined technical problems and assist with the design of systems components or processes to meet specified needs.

PO-4. Engineering Tools, Experimentation and Testing: Apply modern engineering tools and appropriate technique to conduct standard tests and measurements.

PO-5. Engineering practices for society, sustainability and environment: Apply appropriate technology in context of society, sustainability, environment and ethical practices.

PO-6. Project Management: Use engineering management principles individually, as a team member or a leader to manage projects and effectively communicate about well-defined engineering activities.

PO-7. Life-long learning: Ability to analyse individual needs and engage in updating in the context of technological changes.

COURSE OUTCOMES

After competing this course, student will be able to:

CO-1: Interpret the symbols, signs, and conventions from the given drawing.

CO-2: Prepare line plans of residential and public buildings using principles of planning.

CO-3: Prepare submission and working drawing for the given requirement of Load Bearing Structure.

CO-4: Prepare submission and working drawing using CAD for the given requirement of Framed Structure.

CO-5: Draw two-point perspective drawing for given small objects

Mapping of Course Outcomes with Programme Outcomes to be done according to the matrix given below:

Course Outcomes	EXPECTED MAPPING WITH PROGRAMME OUTCOMES (1- Weak Correlation; 2- Medium Correlation; 3- Strong Correlation)						
	PO-1	PO-2	PO-3	PO-4	PO-5	PO-6	PO-7
CO-1	3	1	2	1	2	1	3
CO-2	3	1	3	2	1	1	3
CO-3	3	1	3	2	1	1	3
CO-4	3	2	3	3	1	1	3
CO-5	1	3	3	3	1	1	3

GUIDELINES FOR TEACHERS

To implement Outcome Based Education (OBE) knowledge level and skill set of the students should be enhanced. Teachers should take a major responsibility for the proper implementation of OBE. Some of the responsibilities (not limited to) for the teachers in OBE system may be as follows:

- Within reasonable constraint, they should manoeuvre time to the best advantage of all students.
- They should assess the students only upon certain defined criterion without considering any other potential ineligibility to discriminate them.
- They should try to grow the learning abilities of the students to a certain level before they leave the institute.
- They should try to ensure that all the students are equipped with the quality knowledge as well as competence after they finish their education.
- They should always encourage the students to develop their ultimate performance capabilities.
- They should facilitate and encourage group work and team work to consolidate newer approach.
- They should follow Blooms taxonomy in every part of the assessment.

Bloom's Taxonomy

Level	Teacher should Check	Student should be able to	Possible Mode of Assessment
Creating	Students ability to create	Design or Create	Mini project
Evaluating	Students ability to Justify	Argue or Defend	Assignment
Analysing	Students ability to distinguish	Differentiate or Distinguish	Project/Lab Methodology
Applying	Students ability to use information	Operate or Demonstrate	Technical Presentation/ Demonstration
Understanding	Students ability to explain the ideas	Explain or Classify	Presentation/Seminar
Remembering	Students ability to recall (or remember)	Define or Recall	Quiz

GUIDELINES FOR STUDENTS

Students should take equal responsibility for implementing the OBE. Some of the responsibilities (not limited to) for the students in OBE system are as follows:

- Students should be well aware of each UO before the start of a unit in each and every course.
- Students should be well aware of each CO before the start of the course.
- Students should be well aware of each PO before the start of the programme.
- Students should think critically and reasonably with proper reflection and action.
- Learning of the students should be connected and integrated with practical and real life consequences.
- Students should be well aware of their competency at every level of OBE.

ABBREVIATIONS AND SYMBOLS

List of Abbreviations

Sr. No.	Abbreviation	Full Form
1.	B.B.M	Burnt Brick Masonry
2.	C.M	Cement Mortar
3.	D.P.C	Damp Proof Course
4.	C.R	Coursed Rubble
5.	D	Door
6.	W	Window
7.	V	Ventilator
8.	W.C	Water Closet
9.	F.S.I	Floor Space Index
10.	F.A.R	Floor Area Ratio
11.	T.W	Teak Wood
12.	M.H	Man Hole
13.	N.T	Nahani Trap
14.	G.T	Gully Trap
15.	G. L	Ground Level
16.	C.I	Cast Iron
17.	R.C.C	Reinforced Cement Concrete
18.	P.CC	Plain Cement Concrete
19.	G. I	Galvanized Iron
20.	A.C	Asbestos Sheet

List of Symbols

Sr. No	Electrical Installation	Symbols
1	Light Bracket	
2	Florescent Tube	
3	One way Switch	

Sr. No	Electrical Installation	Symbols
4	Two-way Switch	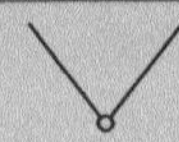
5	Intermediate Switch	
6	Ceiling Fan	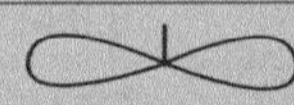
7	Bracket Fan	
8	Exhaust Fan	
9	Fan Regulator	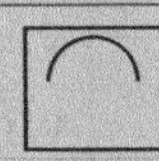
10	Earthing	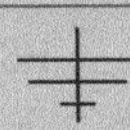
11	2 Pin Socket	
12	3 Pin Socket	
13	Switch Board	
14	Horn or Hooter	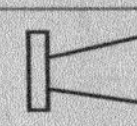
15	Loudspeaker	

LIST OF FIGURES

LIST OF TABLES

CONTENTS

1 Conventions and Symbols

UNIT SPECIFICS

Through this unit we have discussed the following aspects:

- Different Symbols used in drawings for various Materials, Sanitary and Electrical Installations.
- Different types of Lines used in making and reading the Architectural Drawings
- Different size of Lettering and Numerals.
- Different types of Scales used for various types of drawings.
- Types of Paper or Sheets to make drawings.
- To read Drawings of Other Architects.

UNIT OUTCOMES

List of outcomes of this unit is as follows:

U1-O1: Indicate various symbols used in Architectural Drawings

U1-O2: Describe Units and scales in the drawings

U1-O3: Describe various lines in any drawing.

U1-O4: Indicate appropriate size of Lettering.

U1-O5: Apply the details to read and interpret Architectural building drawing.

Unit-1 Outcomes	EXPECTED MAPPING WITH COURSE OUTCOMES (1- Weak Correlation; 2- Medium Correlation; 3- Strong Correlation)				
	CO-1	CO-2	CO-3	CO-4	CO-5
U1-O1	3	3	3	3	1
U1-O2	1	3	3	3	1
U1-O3	1	3	3	3	1
U1-O4	1	3	2	2	1
U1-O5	3	3	3	3	3

1.1 INTRODUCTION

Drawing is the language of Architects and Engineers working in construction and building materials industry. For Civil Engineering, drawings related to construction projects such as Buildings, Bridges,

Atomic Power plants, roads, docks, tunnels etc need to be prepared and submitted. The approved ones, then have to be understood and interpreted for the implementation and execution of the work on site.

The drawings related to residential buildings, commercial buildings and other public buildings come under the category of Building Drawing.

In India, Bureau of Indian Standards (BIS) has laid some standards in building drawings for uniformity and convenience. These norms are known as Conventions of Building Drawings.

1.2 CONVENTIONS AND SYMBOLS

For building drawings, the BIS code **IS 962-1967** is applicable. Conventions as per **IS 962-1967** provide:

1. Scale of drawing and size of lettering.
2. Dimensioning methods, Symbols and abbreviations
3. Methods of Projections.
4. Units of measurements in drawings.

Scan Me

1.2.1 Symbols

Different symbols are used for different construction materials and fixtures as standards in BIS 962 like different building components, water supply and its fixtures, electrical appliances etc.

These *symbols* are *standards* in *every* drawing for the convenience of the workers to read the drawings properly.

Table 1.1: Graphical symbols for different construction materials

Sr. No.	Materials	Symbols
1	Brick work	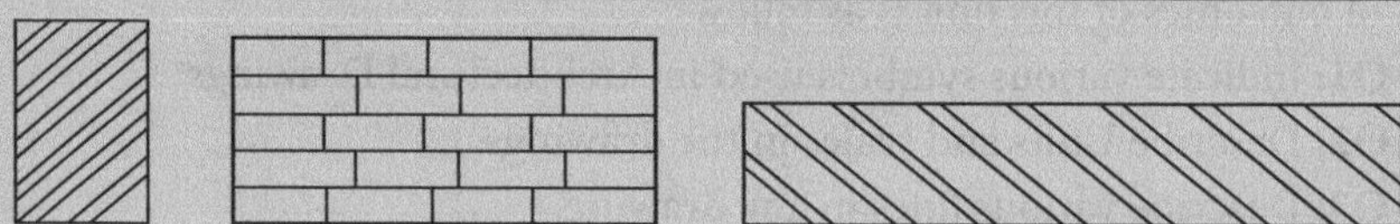 Section Elevation Plan
2	Wood work	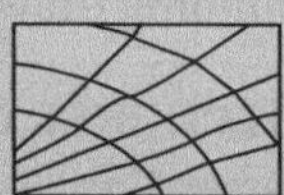
3	Concrete	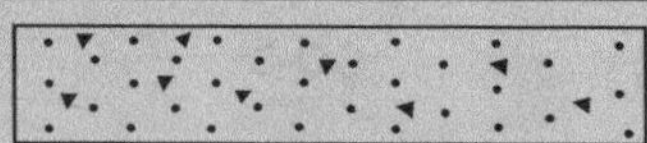
4	Earth work	
5	Glass	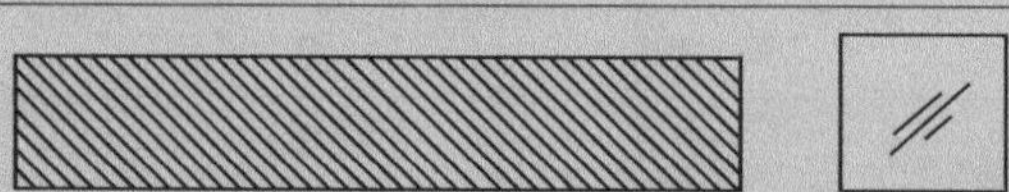 Section Elevation
6	Stone work	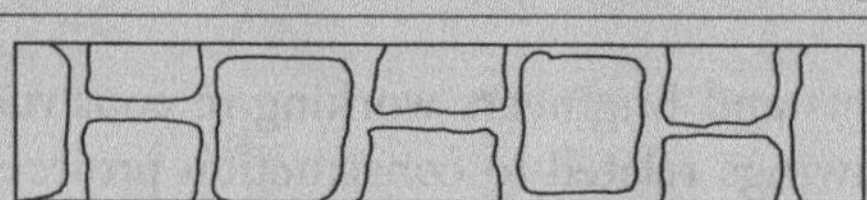

DIPLOMA COURSE BOOKS

S.No.	ISBN	Book Name	Author
37	9789391505516	Engineering Workshop Practice	A. K. Sarathe
38	9789391505653	Environmental Science	Subrat Roy
39	9789355384966	Estimating, Costing and valuation	Sandeep Panchal
40	9789355389510	Fluid Mechanics & Hydraulic Machinery	Suman Chakraborty, Sourav Mitra, Aditya Bandopadhyay
41	9789391505592	Fundamentals of Electrical and Electronics Engineering (with Lab Manual)	Susan S. Mathew, Saji T. Chacko
42	9789355387240	Fundamentals of Power Electronics	Soumitra Kumar Mandal
43	9789355389602	Geotechnical Engineering (Theory & Practicals)	Neelima Satyam
44	9789355386137	Induction, Synchronous, and Special Electrical Machines	K. Selvajyothi, S. Tamilselvi, Sarat Kumar Sahoo
45	9789355382290	Industrial Engineering & Management	A. Manimaran
46	9789355384959	Internet of Things	Sujata Pal
47	9789355388650	Introduction to DBMS: Theory & Practicals	Myneni Madhu Bala
48	9789355384294	Introduction to e-Governance	Kamal Kant Verma
49	9789355386564	Introduction to Electric Generation Systems	Lalit Chandra Saikia, Nalin Behari, Deb Choudhury
50	9789391505585	Introduction to IT Systems (with Lab Manual)	Prashant Joshi
51	9789355382900	Linear Integrated Circuit	B. T. Krishna
52	9789355388148	Manufacturing Engineering	Santosh Kumar
53	9789355383877	Material Science & Engineering	Vinod Kumar, Nisheeth k. Prasad
54	9789391505424	Mathematics-I	Deepak Singh
55	9789391505523	Mathematics-II	Garima Singh
56	9789355385994	Measurements and Metrology	A. Gnanavelbabu, M. Santhana kumar, R. Saranraj
57	9789355388131	Mechanics of Materials and their Testing	R. Velmurugan, Pragathi
58	9789355385048	Microcontroller Applications	Ritula Thakur
59	9789355385062	Microcontrollers and Applications	Santanu Chattopadhyay
60	9789355388100	Mobile and Wireless Communication	Maulin Joshi, Urvashi P. Shukla
61	9789355387790	Production & Operations Management	Mayank Kumar Agrawal
62	9789355388704	Public Health Engineering	Rinku Kumar
63	9789355387455	Python Programming	Rupesh Nasre
64	9789355387189	Renewable Energy Technologies	Sanjay Agrawal, Rajeev K. Mishra
65	9789355387479	Software Engineering	Salman Abdul Moiz
66	9789355389749	Theory of Machines & Mechanisms	G. C. Mohan Kumar
67	9789355387332	Theory of Structures	Arunachalam Subramanian Balu
68	9789355389015	Thermal Engineering - I	V. Raghavan
69	9789355389831	Thermal Engineering - II	Parvesh Antil
70	9789355386779	Transportation Engineering (Theory & Practice)	A.K. Raji, K.K. Babu
71	9789355384287	Water Resource Engineering (Theory & Practice)	Prakash C. Swain, Dillip K. Ghose
72	9789355383051	Web Technologies: Theory & Practical	Raman Dugyala

BOOKS ARE ALSO AVAILABLE IN

HINDI | MARATHI | TAMIL | BENGALI | GUJARATI

Also available as an ebook

For Detailed Catalogue, Please Scan QR Code

AICTE'S LIST OF PRESCRIBED TEXTBOOKS

As Per Model Curriculum & Outcome Based Education

aligned with

NATIONAL EDUCATION POLICY 2020

UNDERGRADUATE COURSE BOOKS

S.No.	ISBN	Book Name	Author
1	9789355382160	Advanced Electrical Machines: Theory and Practicals	Bharat Singh Rajpurohit
2	9789355382139	Analog Electronic Circuits: Theory and Practical	Venu Sonti, Sachin Jain
3	9789355383808	Analog Electronic Devices: Theory and Practicals	N. B Balamurugan
4	9789391505158	Basic Electrical Engineering (with Lab Manual)	S.K. Sahdev
5	9789355388087	Basic of Electronics	Brijesh Iyer
6	9789355387127	Basics of Electrical Machines: Theory and Practicals	D. K. Palwalia, U. K. Kalla, R. K. Kumawat
7	9789355389381	Basics of Mechanics	Shubhankar Bhowmick, N. Venkata Swamy Naidu
8	9789355386250	Basics of Thermodynamics	Parmod Kumar, Atul Dhar
9	9789355386984	Biology for Engineers	Annamma Odaneth
10	9789391505141	Chemistry-I	Manisha Agarwal
11	9789355384843	Civil Engineering Societal and Global Impact	Shakuntala Acharya
12	9789355382184	Computer Aided Civil Engineering Drawing	D. Chandra Mouli, B. Swathi
13	9789355382580	Computer Aided Design & Analysis	Hari Kumar Voruganti
14	9789355387592	Computer Architecture	D. Vaithiyanathan
15	9789355385093	Computer Networks	Hari Prabhat Gupta
16	9789355385307	Construction Engineering & Management	Sparsh Johari
17	9789355388728	Control Systems	S.N. Deepa
18	9789355385420	Data Structure and Algorithms	Ranbir Singh Sanasam
19	9789355386700	Design and Analysis of Algorithms	Hari P. Gupta, Rahul Mishra
20	9789355388872	Digital Electronics and System	Abhishek Bhatt
21	9789355386571	Digital Signal Processing	Sanjeev Sharma
22	9789355383396	Disaster management	Pradeep K. Goyal, Anil K. Gupta
23	9789355385222	Discrete Mathematical Structures	Narendra S. Chaudhari, Sharmila S. P.
24	9789355387868	Effective Technical Communication	Pritee Sharma, Sanchita Verma
25	9789355383389	Electrical Circuit Analysis and Network Theory	Shouri Chatterjee
26	9789355388957	Electromagnetic Waves	Uttam M. Pal, Srijith K.
27	9789355384652	Electronic Communication	Vimal Bhatia
28	9789355388711	Embedded Systems	Santanu Chattopadhyay
29	9789355385147	Energy Science and Engineering	M. Rizwan, Majid Jamil
30	9789355387929	Engineering Economics, Estimation & Costing	Yogesh C. Joshi
31	9789355384669	Engineering Geology	R. N. Khare, Ketan Chourasia
32	9789391505066	Engineering Graphics & Design	Pradeep Jain
33	9789391505097	English (with Lab Manual)	Kulbhushan Kumar
34	9789355385963	Environmental Engineering	Chandra Sekhar Matli
35	9789355388117	Fluid Mechanics	Shreenivas N. Londhe, Pradnya R. Dixit
36	9789355388377	Fluid Mechanics & Fluid Machines	R.P. Saini
37	9789355387103	Geotechnical Engineering	Paramita Bhattacharya

UNDERGRADUATE COURSE BOOKS

S.No.	ISBN	Book Name	Author
38	9789355385130	Hydraulic Engineering	Sayantan Ganguly
39	9789355385550	Hydrology & Water Resources Engineering	Mitthan Lal Kansal, P.K. Agarwal
40	9789355389688	Information Technology Workshop in MATLAB	Hari Prabhat Gupta
41	9789355383235	Instrumentation for Civil Engineering Applications	Ashim Kanti Dey, Manita Das
42	9789355387042	Introduction to Civil Engineering	Manabendra Saharia, Nagendra R. Velaga
43	9789355384805	Introduction to Database Systems	Vikram Singh
44	9789355387509	Machine Element & System Design	Sabah Khan
45	9789355385383	Machine Learning	Himanshu Mittal
46	9789355388414	Manufacturing Automation	Chitragupt Swaroop Chitransh
47	9789355387233	Manufacturing Processes	Dheerendra Kumar Dwivedi
48	9789355389886	Materials Engineering	Dheerendra Kumar Dwivedi
49	9789391505172	Mathematics - I Calculus and Linear Algebra [For Non-Computer Science Engineering Branches]	Reena Garg
50	9789391505271	Mathematics - I Calculus and Linear Algebra [For Computer Science Engineering Branches]	Reena Garg
51	9789355384607	Measurements & Metrology	V.P. Chandramohan
52	9789355383730	Mechanics of Materials	Arvind Vyavahare
53	9789355383358	Mechatronics, Robotics & Control	J. Srinivas
54	9789355385116	Microprocessor and Microcontroller	Saurabh Chaudhury, Risha Mal
55	9789355386083	Microprocessors	Yogendra Kumar Gupta
56	9789355389169	Mobile Communication and Networks	Vimal Bhatia
57	9789355385475	Operating Systems	Sukomal Pal
58	9789355385604	Operating Systems	Sanjaya Kumar Panda
59	9789391505134	Physics (Oscillations, Waves and Optics) With lab manual	A.B. Bhattacharya, Atanu Nag
60	9789391505257	Physics (Quantum Mechanics for Engineers) With lab manual	A.B. Bhattacharya, Atanu Nag
61	9789391505165	Physics (Introduction to Electromagnetic Theory) (with Lab Manual)	A.B. Bhattacharya, Atanu Nag
62	9789391505059	Physics (Introduction to Mechanics) (with Lab Manual)	A.B. Bhattacharya, Atanu Nag
63	9789355385956	Power Electronics: Theory and Practicals	Lalit Chandra Saikia
64	9789355384942	Power Systems-I	Suresh Mikkili
65	9789355386236	Power Systems-II	Arun Rahul S, Manas K. Jena
66	9789355383754	Product Innovation & Entrepreneurship	Vigneswaran C
67	9789355384584	Production & Operation Management	Navin Kumar Dev
68	9789391505219	Programming for Problem Solving (with Lab Manual)	R.S. Salaria
69	9789355385451	Signals and Systems	Sanjay L. Nalbalwar
70	9789355386021	Strength of Materials	Uday Shanker Dixit, Nelson Muthu, S. M. Kamal
71	9789355383792	Structural Engineering	T.D. Gunneswara Rao, D. Ravi Prasad
72	9789355387004	Surveying and Geomatics	P. K. Garg
73	9789355384317	Testing & Evaluation of Civil Engineering Materials	Sandeep Chaudhary, Kashyap A Patel
74	9789355382306	Theory of Computation	Debi Prasanna Acharjya
75	9789355384928	Transportation Engineering	A.K. Raji
76	9789355385208	VLSI Design	Mahendra Sakare, Shonal Chouksey
77	9789391505332	Workshop / Manufacturing Practices (with Lab Manual)	Veeranna D.K.

DIPLOMA COURSE BOOKS

S.No.	ISBN	Book Name	Author
1	9789355382481	Aanandam An Act of Community Service	Shuchi Sharma
2	9789355385567	Advanced Manufacturing Processes	Ranjeet Kumar Sahu, Devendra Laxman Kamble
3	9789355385109	Advanced Surveying: Theory & Practice	Ramakant Agrawal, Parshottam Sarathe
4	9789355384546	Algorithms	Piyoosh P, Arnab Sarkar
5	9789391505448	Applied Chemistry (with Lab Manual)	Anju Rawlley, Devdatta Vinayakrao Saraf
6	9789391505431	Applied Physics-I (with Lab Manual)	Vinod Kumar Yadav, Mina Talati
7	9789391505578	Applied Physics-II (with Lab Manual)	Hussain Jeevakhan
8	9789355387530	Basic Mechanical Engineering	Dharmendra Singh, Chandra Shekhar Rajoria
9	9789355387394	Building Construction	Rinku Kumar, Sandeep Panchal
10	9789355383280	Building Electrification	Arnab Ghosh
11	9789355384379	Building Planning and Drawing	Aparna Dwivedi
12	9789391505493	Communication Skills in English	Anjana Tiwari
13	9789355383570	Computer Aided Machine Drawing Practice	Kanak Kalita
14	9789355388964	Computer Networking and Data Communication	Sanjaya Shankar Tripathy
15	9789355382986	Computer Networks: Theory & Practicals	Brijendra Pratap Singh, Manoj Madhava Gore
16	9789355382672	Computer Programming: Theory and Practicals	Satyendra Singh Chouhan
17	9789355387714	Computer System Organization	Sonal Yadav
18	9789355382658	Concrete Technology: Theory and Practice	Sasankasekhar Mandal, Vishwajit Anand, Sushil K. Agarwala
19	9789355383334	Construction Materials and Testing	Vanita Aggarwal
20	9789355383501	Consumer Electronics	Amit M. Joshi, Maulin M. Joshi, Urvashi Prakash Shukla
21	9789355387684	Data Structures: Theory & Practicals	NB Venkateswarlu
22	9789355388513	Design of Machine Elements	A. Kumaravel, M. Kathirselvam
23	9789355388360	Design of Steel and RCC Structure	Sparsh Johari
24	9789355383303	Digital Communication	Muralidhar Kulkarni, K. S. Shivaprakasha
25	9789355387295	Digital Electronics	Menka Yadav
26	9789355389435	Electric Circuits and Networks	Jamuna K., Nilanjan Tewari
27	9789355388452	Electric Motors and Transformers (Theory and Practicals)	Sanjay B. Bodkhe
28	9789355385901	Electric Power Transmission and Distribution	Sudhir Sharma
29	9789355389466	Electrical & Electronics Measurements and Instrumentation	Sudarsan Sahoo, Vipin Chandra Pal, Sudipta Chakraborty
30	9789355382696	Electronic Communication	M. D. Selvaraj, P. Nagaradjane
31	9789355387202	Electronic Devices and Circuits	Ankesh Jain
32	9789355386373	Electronic Equipment Maintenance	Chanchal Sharma
33	9789355384690	Embedded Systems	Deepa M, G. Santhanamari, Susithra. N
34	9789355384867	Energy Conservation and Audit	M. Rizwan, Majid Jamil
35	9789391505509	Engineering Graphics	Sharad K. Pradhan, K.K. Jain
36	9789391505608	Engineering Mechanics (with Lab Manual)	Bhankhar Bharat, Gokaldas Vandana Somkuwar

1.3 GRAPHICAL SYMBOLS FOR DOORS AND WINDOWS

1.3.1 Windows

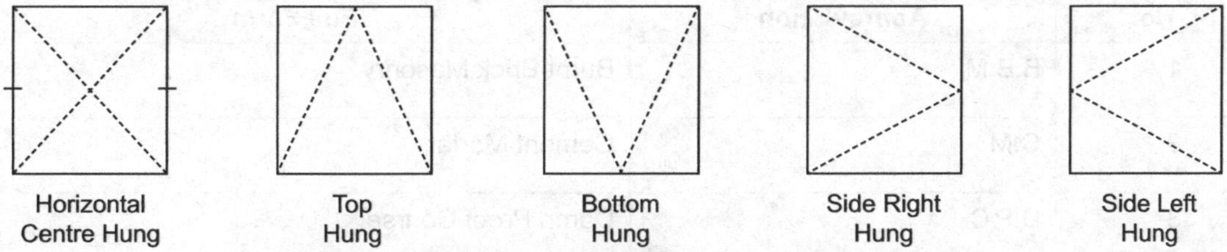

Fig 1.1: Graphical symbols for different windows

1.3.2 Different Types of Doors

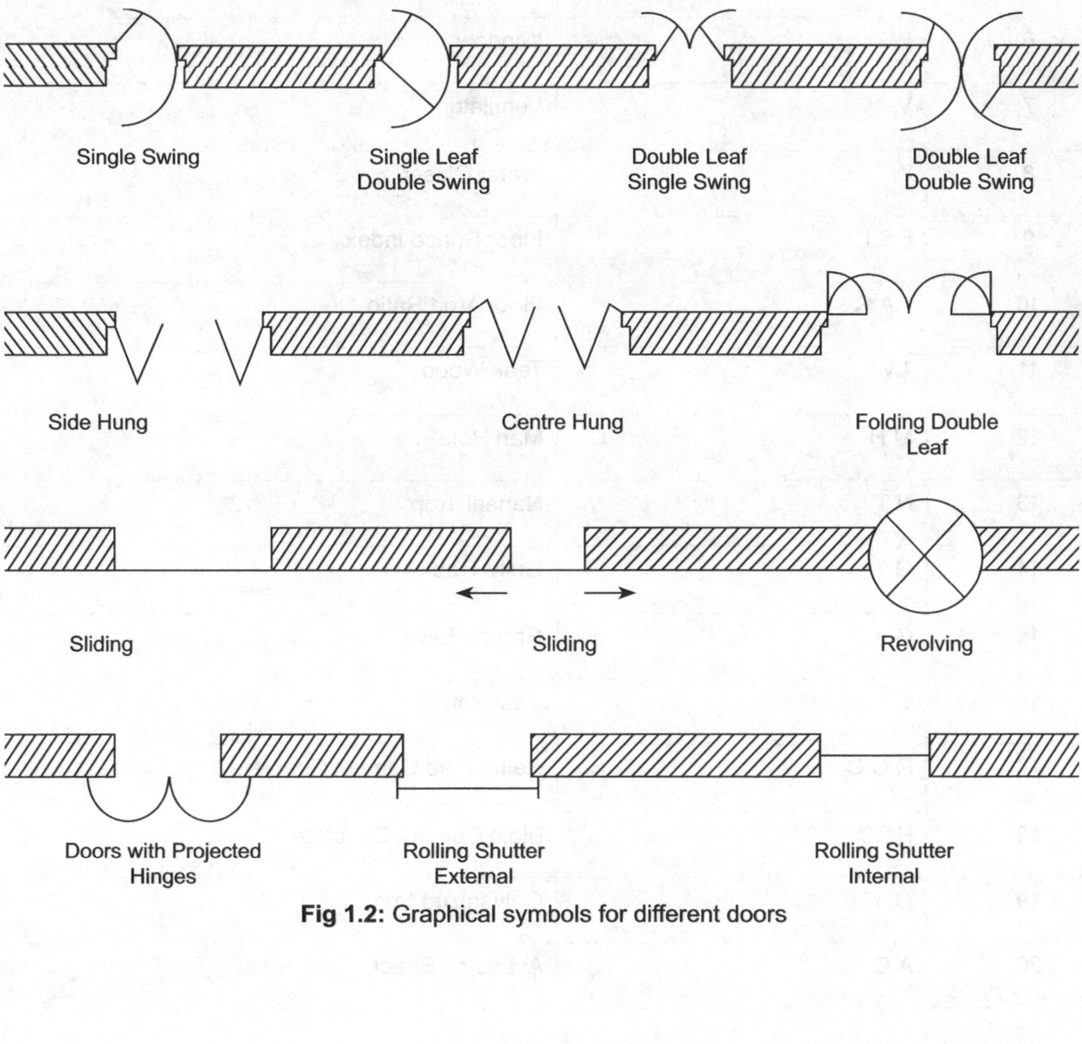

Fig 1.2: Graphical symbols for different doors

1.3.3 Abbreviations

Any drawing of building includes all components with their dimensions and materials of construction.

Table 1.2: Abbreviations Used in the drawings

Sr. No.	Abbreviation	Full Form
1	B.B.M	Burnt Brick Masonry
2	C.M	Cement Mortar
3	D.P.C	Damp Proof Course
4	C.R	Coursed Rubble
5	D	Door
6	W	Window
7	V	Ventilator
8	W.C	Water Closet
9	F.S.I	Floor Space Index
10	F.A.R	Floor Area Ratio
11	T.W	Teak Wood
12	M.H	Man Hole
13	N.T	Nahani Trap
14	G.T	Gully Trap
15	G. L	Ground Level
16	C.I	Cast Iron
17	R.C.C	Reinforced Cement Concrete
18	P.CC	Plain Cement Concrete
19	G. I	Galvanized Iron
20	A.C	Asbestos Sheet

1.3.4 Electrical Installation Symbols

Table 1.3: Graphical symbols for electrical installations

Sr. No	Electrical Installation	Symbols
1	Light Bracket	
2	Florescent Tube	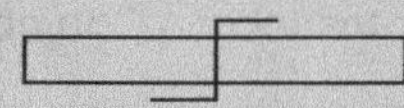
3	One way Switch	
4	Two-way Switch	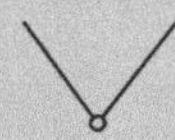
5	Intermediate Switch	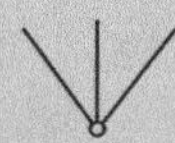
6	Ceiling Fan	
7	Bracket Fan	
8	Exhaust Fan	
9	Fan Regulator	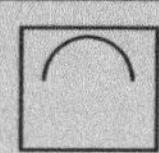
10	Earthing	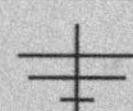
11	2 Pin Socket	
12	3 Pin Socket	
13	Switch Board	
14	Horn or Hooter	
15	Loudspeaker	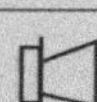

1.3.5 Symbols for Sanitary Installations

Table 1.4: Graphical symbols for Sanitary and other useful installations

Sr. No.	Object	Symbol
1	Gully	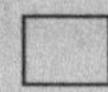
2	Man hole or Inspection Chamber	
3	Intercepting Trap and Fresh Air Inlet	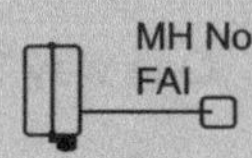
4	Drain Cock	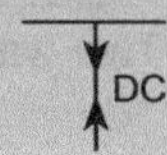
5	Plain Kitchen Sink	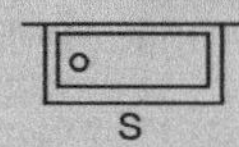
6	Double Kitchen Sink	
7	Refrigerator	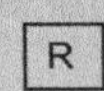
8	Shower Stall	
9	Shower Head	
10	WC Indian Type	
11	Towel Rail	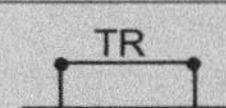
12	Rain Water Gutter	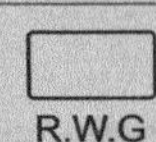
13	Mixing Valve	
14	Water Meter	
15	Fire Extinguisher	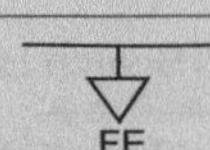
16	Fire Hydrant	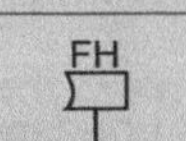
17	Stairs	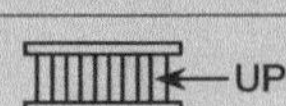
18	Bed	

Sr. No.	Object	Symbol
19	Stop Valve	
20	Safety Valve	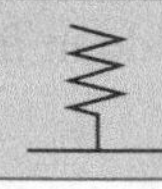

1.4 TYPES OF LINES

There are different types of lines in any drawing. Every line has a different meaning in drawing. As an engineer we need to read drawings and according to the drawing the work on the site should be executed. Various lines which you will see in drawings are indicated below:

1.4.1 Types of Lines

1. Visible Out Lines
2. Centre Lines
3. Hidden Lines
4. Section Lines
5. Dimension Lines
6. Extension Lines
7. Pointer Lines

1.4.1.1 Visible Out Lines

The visible out lines should be thick outlines according to the purpose, it can be 0.60,0.80, 1.00 or 1.30mm. These lines should be outstanding in appearance. These lines are used to show wall thickness, plot boundary, proposed structure etc. Thickness may be 0.6mm to1.3mm.

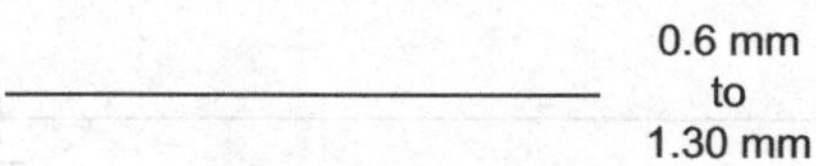

Fig 1.3: Visible Out Lines

1.4.1.2 Centre Lines

These lines are marked on the drawings to show centre of component it can be column, beam and wall. Thickness 0.2mm to 0.3mm.

These can be extended to the outlines to write dimensions and it can be long and short lines and be in the ratio of 4:1 or 6:1 closely and evenly spaces out. In any drawing the ratio one adopted should be maintained.

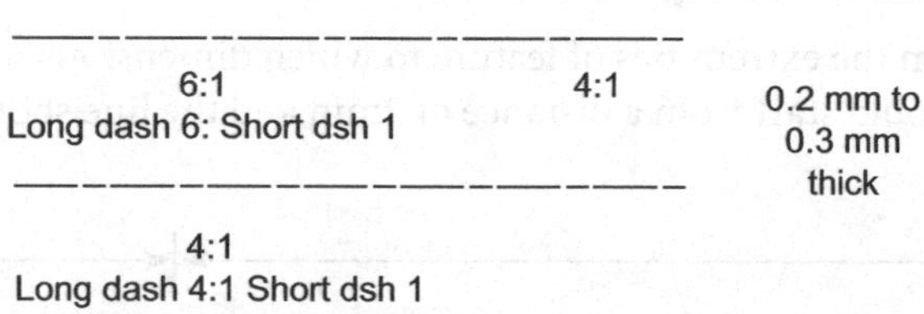

Fig 1.4: Centre Lines

1.4.1.3 Hidden Lines

These lines are shown when there is a hidden surface or something is not visible on the surface which is drawn. These have to be shown in close and evenly spaced dashes, thickness 0.4mm-0.5mm. Should be used to represent hidden lines. These lines in the drawings to show Loft, R.C.C Chajja, Valley, Canopy etc.

Fig 1.5: Hidden Lines

1.4.1.4 Section Lines

The plane on which the section has to be taken section lines should be indicated in long dash and two short dash alternately and evenly spaced out also arrow to be shown in the direction of the section to see and name of the section at the end in bold letters. Thickness 0.6mm to 1.0mm.

Fig 1.6: Section Lines

1.4.1.5 Dimension Lines:

Thin firm line is used in contrast with heavier visible outlines to show dimension of the structure. Thickness is 0.2mm – 0.3mm. the drawing is complete only when dimensions are correctly mention on the drawing with correct placement.

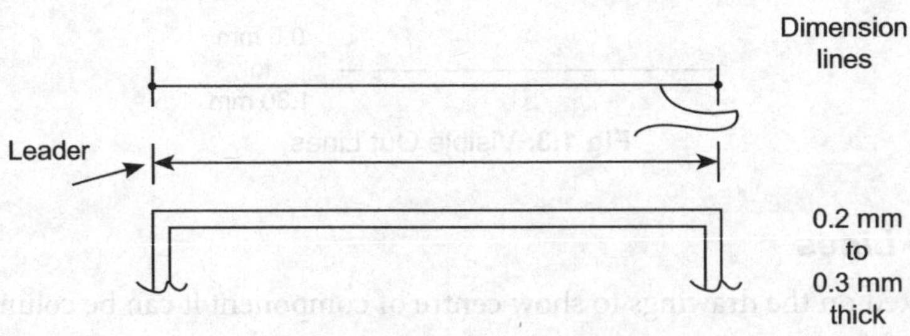

Fig 1.7: Dimension Lines

1.4.1.6 Extension Line OR Light and thin lines

These lines are drawn from the extremities of feature to which dimensions has to be given. It should not touch the feature but it should start from a distance of 2mm and the line should extend till 4mm beyond the dimension line.

Fig 1.8: Extension Line

1.4.1.7 *Pointer Line*

When space is insufficient to write a note or dimension near the feature, pointer lines are drawn from the note or dimension to show where it applies. It can be known as leader or pointer. To point out the names or the figure these pointers are used.

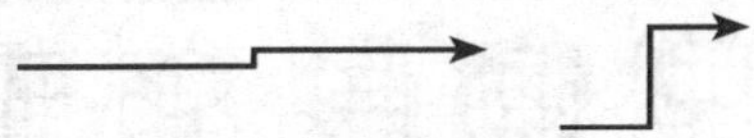

Fig 1.9: Pointer Line

1.4.1.8 *Break Line OR Short break lines*

These are free hand drawn lines. To show continuity of the object, long break line or short break line is used. Suppose a wall is continued bet the paper is not that big, we can show break lines.

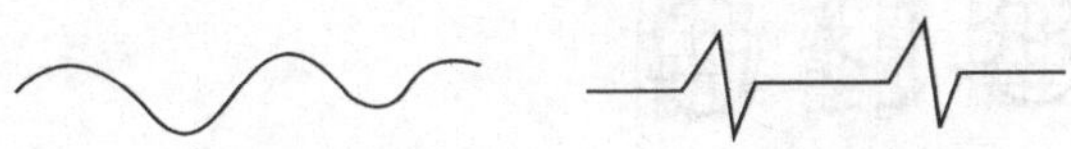

Short Break Long Break

Fig 1.10: Break Line

1.4.1.9 *Arrow and Dots*

These are at the ends of dimension lines. The length of the arrow is about 4 times the depth. The space of the arrow should be filled in. in case of dot it should be placed between dot dashed and dimension line. The sharp pencil should be used to make these neat arrows.

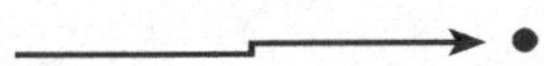

Fig 1.11: Arrow and Dots

1.4.2 Appropriate Size of Lettering and Numerals for Title, Subtitles, Notes and Dimensions

It is a very important to write clean and uniform letters on the drawings. All letters should be capital and must have standard shape and size as shown in table

Scan Me

Table 1.5: Size of Lettering and Numerals

Sr. No.	Purpose	Size of Lettering and numerals in mm
1.	Main Title and Drawing	6,8,10 and 12
2.	Subtitles and Headings	3,4,5 and 5
3.	Notes, schedule , Material and dimension	2,3,4 and 5

A B C D E F G H I J K L M N O P Q R S T U V W X Y Z 1 2 3 4 5 6 7 8 9 10

A B C D E F G H I J K L M N O P Q R S T U V W X Y Z 1 2 3 4 5 6 7 8 9 10

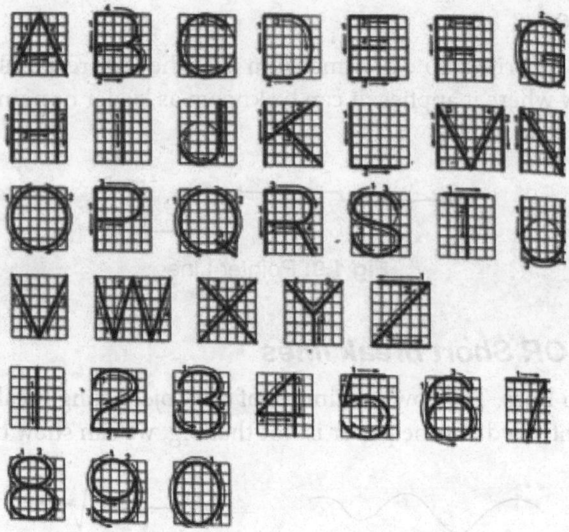

Fig 1.12: Standard shapes of letters

1.5 TYPES OF SCALE

Definition of Scale: Scale is a tool to measure the distance and transferring measurements at a fixed ratio of length.

In Architecture, scale means relation and proportion of different elements with each other. It makes the size and proportion of the elements like doors, windows, staircases in residential buildings in relation to that of buildings

1.5.1 Intimate Scale

This scale is in relation to human or more personal scale.

1.5.2 Monumental Scale

This is much larger than human scale and is intended to be impressive, e.g., public buildings, memorials, religious buildings, and so on. Proportion: This refers to the relative size of parts of a whole, the relationship between two things of different size.

1.5.3 Criteria for Proper Selection of Scale

Scales indicated below for the preparation of different types of drawings are given:

Table 1.6: Scales used in drawings

Scale	Purpose
1:200, 1:500 or 1:1000	For site plan
1:50 or 1:100	For Developed plan Foundation Plan, Sections
1:100(10 mm = 1m)	For Municipal drawings submission
1:20(50 mm = 1m)	For Details and General Drawing

Scale	Purpose
1:10(100 mm = 1m)	For Enlarge Details
1:5(200 mm = 1m)	For Detail Drawings
1:2 (500 mm = 1m)	For Detail Drawing
1:1(Full Size)	For Detail Drawings

1.6 SIZES OF VARIOUS STANDARD PAPERS /SHEETS

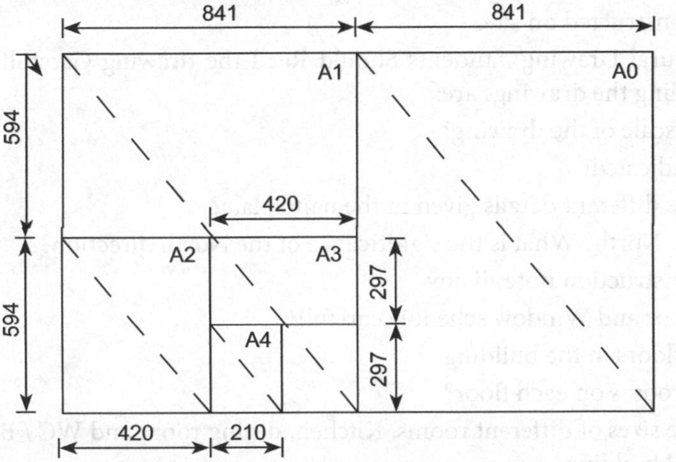

Fig 1.13: Different sizes of standard papers and sheets

International Standards Organization

Table 1.7: Standard paper and sheets

Sheet Designation	Sheet Size
A0	841 x 118
A1	841 x 594
A2	420 x 594
A3	420 x 297
A4	210 x 297

1.7 READING AND INTERPRETING READYMADE ARCHITECTURAL BUILDING DRAWING

1.7.1 Reading a Plan

A Plan is drawing of any object in two dimensions as if viewed from the top. For a building or a room, it gives length and breadth of the space, thickness of the walls, columns, steps, sit outs, width of the door and windows opening, furniture layout and levels of the ground.

1.7.2 Reading an Elevation

An Elevation is the face of the building used for presentation and for aesthetic purposes. One can see the height of the building, height of the doors, windows, level of the chajja and other projections, height of the verandah, levels of the steps etc. Dimensions are not to be written in Elevation.

1.7.3 Reading a Perspective

It is the presentation drawing in three dimensions. Through perspective the actual height of the building, the elements in the building can be seen clearly with proportions. Effect of painting in different colours can be visualized before constructing the building. Perspective is like actual view of the building on paper before it gets constructed on site.

In any Architectural Drawing, Students Should Read the drawing Carefully Some points to be considered while reading the drawings are:

1. What is the scale of the drawing?
2. What is it indicated?
3. What are the different details given in the nameplate?
4. Where is the North? What is the significance of the North direction?
5. Read the construction notes if any
6. Read the Door and Window schedule carefully.
7. How many floors in the building?
8. How many rooms on each floor?
9. What are the sizes of different rooms, Kitchen, dining room and WC / Bath and toilets in case of residential building?
10. Study the Elevation, what are the details in Elevation?
11. Study the Section, co-relate the section with the plan. See the Section Line in the plan from one end to the other end of the plan. Observe how section is drawn wall by wall and every feature.
12. How is the wall continuing in the foundation and how column is continuing in the section?
13. Observe the dimensioning, does the dimension shows the actual measurements.
14. Observe the hidden lines and centre lines in the plan and find out the purpose of these lines.
15. Observe the placement of the titles and size of the fonts or lettering.
16. Observe the Layout of the Plan, Elevation, Section, Foundation Plan, Location Plan, Foundation Plan, Site Plan on the Drawing.

Scan Me

While reading, the drawing for any Load bearing structure all details should be read carefully.

1. Depth and width of the foundation trench of the for the wall
2. Thickness of masonry and material used for construction
3. Height and width of all the openings like doors, windows etc.
4. Sill height, lintel height, height of ceiling and Floor thickness is to be read from sections.
5. Thickness of compound walls, Parapet wall, partition wall should be read and consider while construction.

While reading, the drawing for any frame structure all details should be read carefully.

1. Size of the column, number of columns with the size and type of their footings with depth.
2. Sizes of beams, including plinth beams, floor beams.

3. Details of reinforcement in slabs, column, beams and footings
4. Thickness of compound walls, Parapet wall, partition wall should be read and consider while construction.

EXERCISE

Multiple Choice Questions

1. Scale used to make Site Plan
   (a) 1:10 (b) 1:20
   (c) 1:50 (d) 1:500
2. Scale used to make building Plan
   (a) 1:10 (b) 1:20
   (c) 1:50 (d) 1:100
3. Scale used to make details of the door
   (a) 1:10 (b) 1:20
   (c) 1:50 (d) 1:100
4. Hidden Lines on drawings are used for
   (a) Chajja (b) Steps
   (c) Walls (d) Windows
5. Long and Short lines with the ratio of 4:1 or 6:1
   (a) Dimension Lines (b) Centre line
   (c) Extension line (d) Hidden line

Answer to Multiple-Choice Questions				
1. (d)	2. (d)	3. (a)	4. (a)	5. (b)

Long and Short Answers

Category I

1. List different scales use to prepare building drawings
2. List out sanitary installation for toilets
3. List various electrical installations in a residential building.
4. List various types of doors openings
5. Make sketches of:
   (i) Mortar
   (ii) Brickwork in Plan and Section
   (iii) Woodwork
   (iv) Glass
   (v) Stonework

Category II

1. Draw and elaborate upon any four lines as per IS 962.
2. Draw different types of Scales and their importance in drawings.
3. Draw symbols of 5 appliances used in the kitchen as per IS 962.
4. Draw symbols used for Brick work, Woodwork, Concrete, Stone work and Glass as per IS 962.
5. What points are to be consider while reading any Architectural drawing?
6. List various lines which you see in given drawing.
7. Make the alphabets in different sizes on the Graph Paper.
8. List various paper sizes appropriate for drawing purpose.
9. List various symbols used in the toilet design.
10. List various elements on Elevation of the building.

REFERENCES

1. S.V Deodhar, Building Drawing, New Vrinda Publishing House, 1992.
2. Francis dk, Ching, Architectural Graphics, John Willey &Sons Inc, 2015.
3. Ramsey/Sleeper, ARCHITECTURAL GRAPHIC STANDARDS, John Willey &Sons Inc, 2000.
4. N.D Bhatt, Engineering Drawing Graphics, Charotar Publishing House Pvt. Ltd, 2011.
5. https://www.youtube.com/watch?v=P544Koz5SyQ

2 Planning of Building

UNIT SPECIFICS

Through this unit we have discussed the following aspects:

- Different planning principles applied in the building while planning.
- Different principles of planning.
- Different types of scales for various types of drawings.
- Byelaws to be followed while making the plan of any building.
- To make and read drawings of other Architects.
- Different types of line plans to reading the architectural drawings

UNIT OUTCOMES

List of outcomes of this unit is as follows:

U2-O1: Various Principles of planning for residential and public building.
U2-O2: Describe space requirements and norms to design buildings
U2-O3: Rules and byelaws to follow for making drawings.
U2-O4: Different terms used in construction of a building.
U2-O5: Line plan showing the circulation and planning of various Architectural projects.

Unit-2 Outcomes	EXPECTED MAPPING WITH COURSE OUTCOMES (1- Weak Correlation; 2- Medium Correlation; 3- Strong Correlation)				
	CO-1	CO-2	CO-3	CO-4	CO-5
U2-O1	2	1	1	2	1
U2-O2	1	2	3	3	2
U2-O3	2	3	3	1	2
U2-O4	1	1	2	1	-
U2-O5	3	1	1	1	3

2.1 INTRODUCTION

A building is an idea which is in the mind then to give shape and to show others it must be drawn on paper or computer before construction of that building will start. Each building requires different type of planning and treatment as per their function. Building planning is a rigorous planning process. There are many aspects to be considered before planning any building. Requirements of every building is different. To plan any building like residential, commercial or any public building certain principles are followed, these are known as principles of planning. These principles ensure proper planning in utilization of space, orientation of the building, circulation in the building, furniture requirement sanitation, water supply, electrical, and also consider economical aspect. The problem of planning a building can be dealt in two stages:

1. Problem - seeking stage i.e., Programming
2. Problem- solution stage i.e., Real design.

Problem seeking stage i.e. Programming deals to establish goals – to seek the objectives and constraints of the elements like plot size and its location, the soil condition, any tree on the site to retain.

To collect original data and analyse the facts.

- To foresee the future expectations and its development and details.
- To determine real need: what is the history of the project and for future extension if any, what all provisions to be made.
- To state the problem: so, after all the data about the project is known, its constraints, future demands then decide the problem in detail.

  After the problem explained in detail, the actual design of the building starts which is solution stage

  Elaborate problem: Architects need to find out dimensions of each element of the building to design.
- To analyze the existing facilities in that area like water supply, sanitary, electric lines, telephone connection, gas connection, etc and to visit the site to get actual condition of the site.
- To develop space program i.e to start design on the paper. Some alternatives need to be sketched and then choose the best option amongst all and then discuss with the client and his expectations need to be considered. And both the party come on the conclusion after discussion, final drawings to be made like plan. Elevation, section, plumbing details, electrical wiring, fittings need to be plan and drawn on the paper. And then structural members are designed to take the load of the building and the cost of the project will be work out. According to the designs the quantities of materials calculated, and a final report is made.

The points to be considered are:

- Economy and engineering standards.
- Aesthetics and visual features
- Functional planning and treatment.
- Soil investigation results, soil type, strata etc.
- Parking place for 2-wheeler, wheeler etc.
- As per the finalized plan services like lifts, acoustics, fire etc. needs to plan.

After these stages the stage is

- Contract documents
- Construction
- Occupancy/handover
- Feedbacks

2.1.0 Principles of Planning

Functional planning means design the building for functional use. Every building has different functions to be fulfilled. Even single building caters to different function. Like function of an educational building is totally different than a commercial complex. Then the function of the toilet is totally different than a kitchen. So different buildings have their own specific function. That is why their requirement are also different.

2.1.1 Aspect

It is defined as the direction through which the room receives natural light and ventilation from the surrounding.

ike to take maximum advantage of sunlight and wind, planning is required in positioning of windows and doors on the external walls different rooms in the building have different requirements and function. To plan these the aspect changes according to the room.

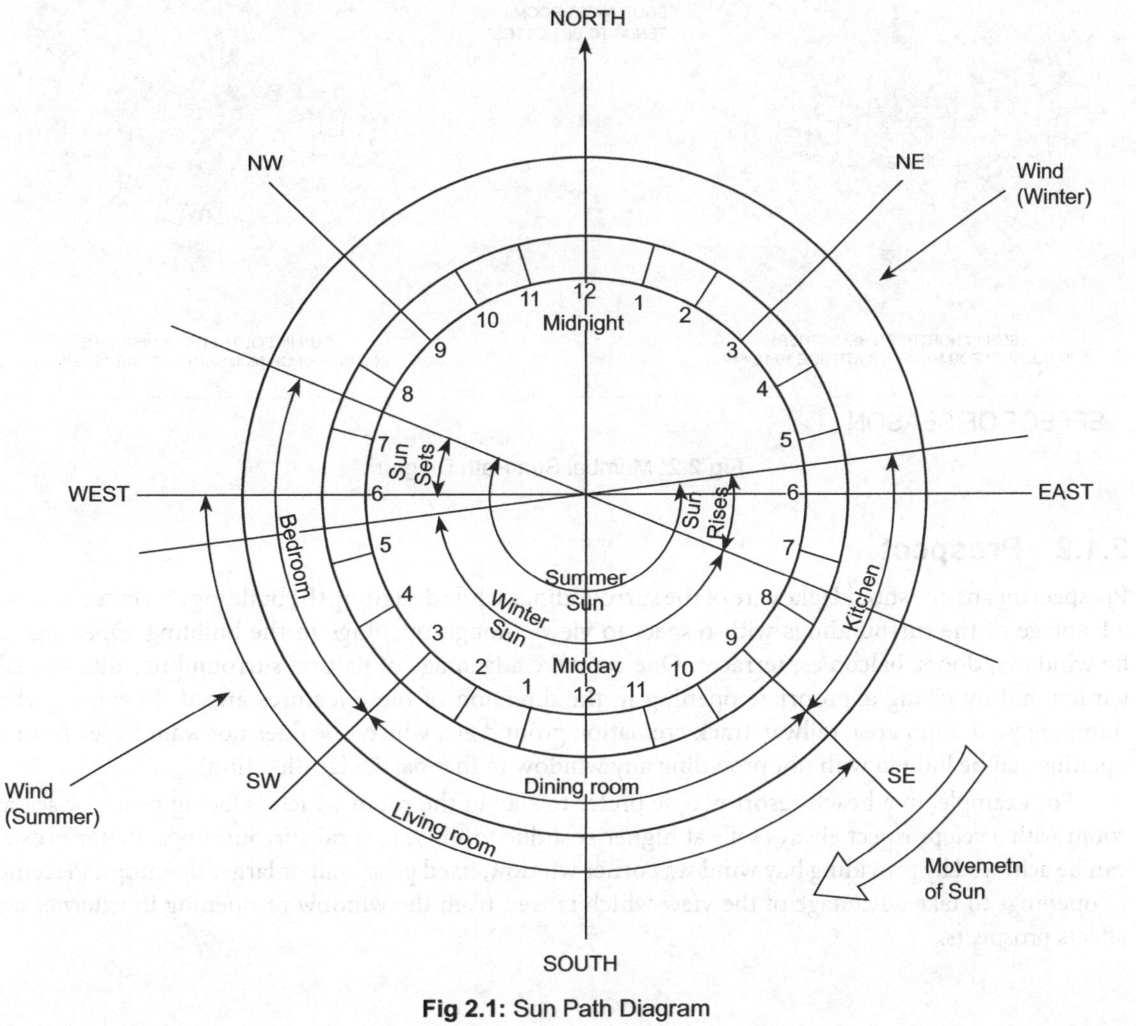

Fig 2.1: Sun Path Diagram

Like kitchen should have one opening in the east direction to take advantage of the morning sun rays. morning sunrays has UV rays which kills germs, this aspect is very essential to consider while planning a kitchen. In this way if planning a toilet, it should be planned in such a way that it will block the heat from the west direction or south direction. The living room should face southern or southeast aspect because the sun is towards south during cooler days and with the southern aspect it will have sunshine. In winter and less sunshine in summer. All the bedrooms may have west or southwest aspect because the wind direction in summer would be from this particular side. And a verandah, balcony or gallery can be planned to protect the outer wall from harsh sun and rain. Generally, while planning the classrooms in the school, north light aspect to be considered to take advantage to the diffused and evenly distributed light.

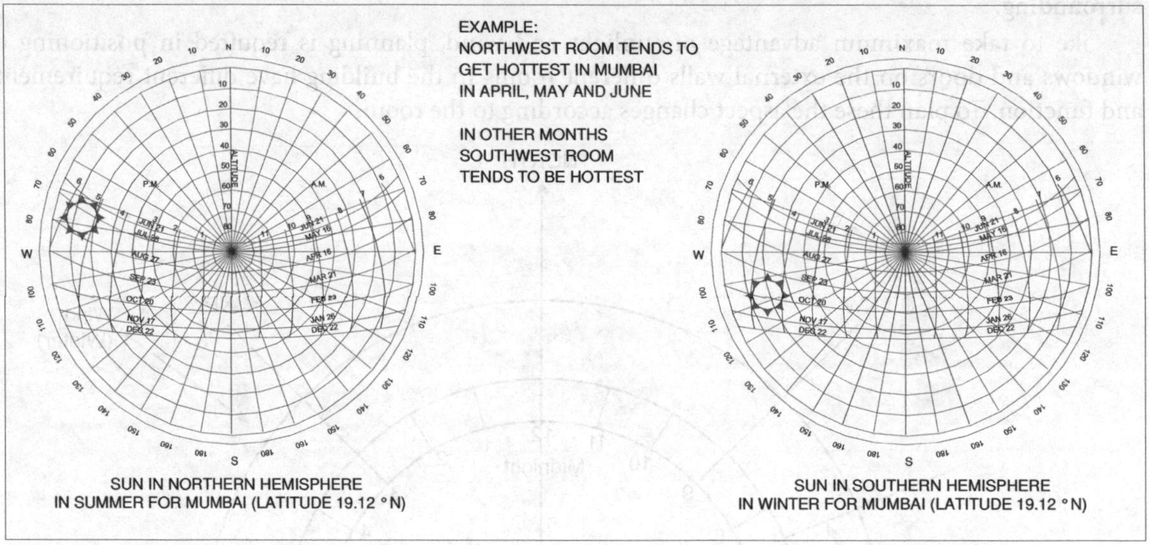

EFFECT OF SEASON

Fig 2.2: Mumbai Sun Path Diagram

2.1.2 Prospect

Prospect means one should take care of the surroundings while designing the building and take maximum advantage of the surroundings with respect to views though openings in the building. Openings can be windows, doors, balconies, terraces. One can take advantage of natural surroundings like sea, lake, garden, hill by giving appropriate opening in the direction of these features and if there are garbage dumping yard, slum area, railway track cremation ground etc. which one does not want to see from the opening can be hidden with not providing any window in that particular direction.

For example, in a beach resort people prefer to stay in the room which is facing beach or sea and room with a sea prospect always sells at higher cost due to those natural surroundings. Better prospect can be achieved by providing bay window, corner window, fixed glass wall or larger openings. Placement of openings to take advantage of the view which is seen from the window or opening in external walls affects prospects.

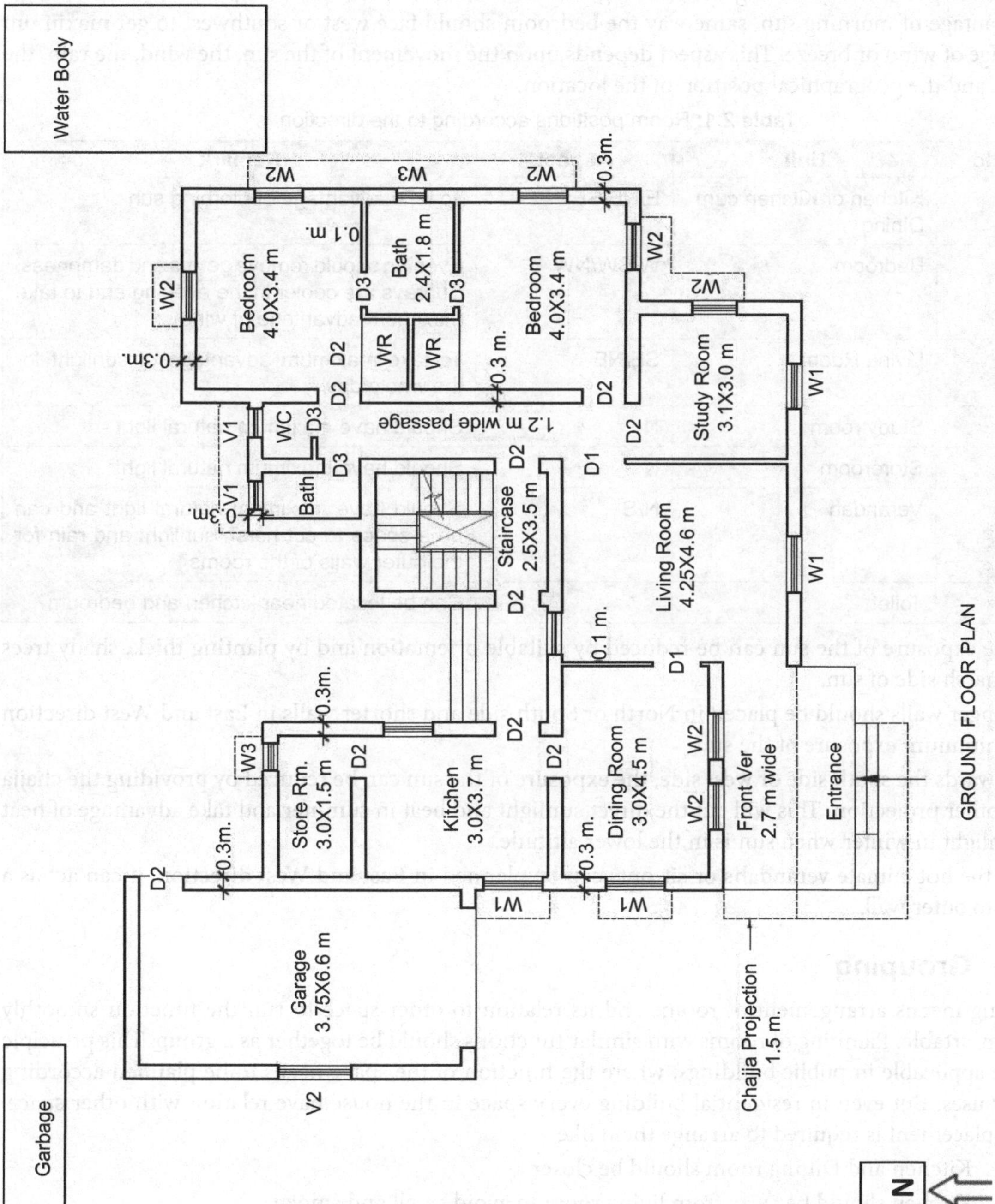

GROUND FLOOR PLAN

Fig 2.3: Window positions prospect

2.1.3 Orientation

Orientation means to place the room or building with respect to North direction. Directions are very important aspect while planning any building. Like kitchen should have opening in the east wall, to get the advantage of morning sun, same way the bedroom should face west or southwest to get maximum advantage of wind or breeze. This aspect depends upon the movement of the sun, the wind, the rain, the climate, and the geographical position of the location.

Table 2.1: Room positions according to the direction

Sr. No	Unit	Aspect	Remark
1.	Kitchen or Kitchen cum Dining	E/NE/SE	To take advantage of Morning sun
2.	Bedroom	W/SW/NW	Evening should remove germs and dampness. Sunrays are cooler in the evening and to take maximum advantage of wind.
3	Living Room	SE/NE	To take maximum advantage of sunlight for the whole day.
4.	Study room	N	Should have maximum natural light
5	Storeroom	N	Should have maximum natural light
6	Verandah	N/S	Should have maximum natural light and can be a space to cut harsh sunlight and rain for the outer walls of the rooms
7	Toilet		Can be located near kitchen and bedroom

The exposure of the sun can be reduced by suitable orientation and by planting thick, shady trees on the harsh side of sun.

Longer walls should be placed in North or South side and shorter walls in East and West direction to get minimum exposure of the sun.

Towards the south side or west side, the exposure of the sun can be reduced by providing the chajja or any other projection. This will cut the direct sunlight and heat in summer and take advantage of heat and sunlight in winter when sun is in the lower altitude.

In the hot climate verandahs or sit-outs can be planned in East and West direction, it can act as a barrier to outer wall.

2.1.4 Grouping

Grouping means arrangement of rooms and its relation to other space to run the function smoothly and comfortable. Planning of rooms with similar functions should be together as a group. This principle is more applicable in public buildings, where the function of the space needs to be planned according to their uses. But even in residential building every space in the house have relation with other space, proper placement is required to arrange them like

1. Kitchen and Dining room should be closer
2. Kitchen should be away from living room to avoid smell and smoke.
3. Kitchen should be closer to entrance of the building and also to the toilets. kitchen should be group with toilets or WC and Bath to facilitate water supply and drainage.

4. Staircase should be centrally located for easy approach from all the rooms.
5. In hospitals, administrative area, wards, operation theatres etc should be grouped according to the function.

In planning as a whole grouping of similar spaces plays a very important role, it creates an internal harmony.

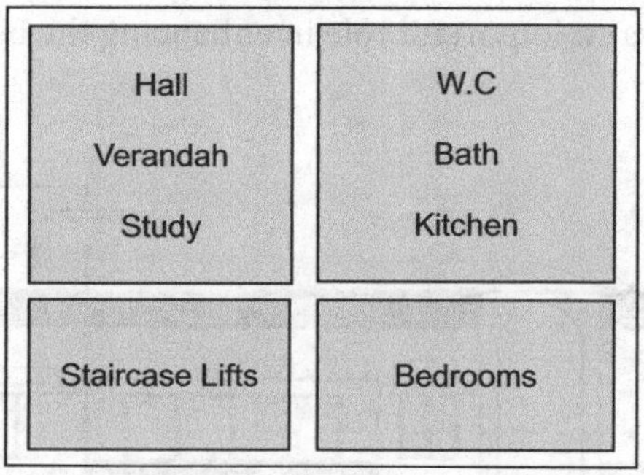

Fig 2.4: Residential building Groups

2.1.5 Privacy

Privacy is an important principle while planning all types of buildings whether public or residential. Privacy means segregating one function from another. It can be external or internal. As an example, maintaining distance from the nearby building and/or public street is an external privacy while for a library the reading room and study room may be separated from the main function of the library is internal privacy.

There are two types of privacy

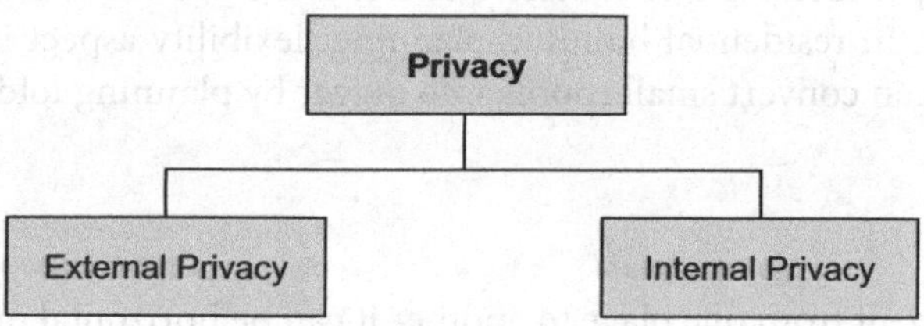

Fig 2.5: Types of Privacy

(a) ***External privacy***: It means privacy from surrounding buildings as well as privacy from noise and pollution from the road. By good planning external privacy can be maintained like depending upon the location of the opening, the heights of the sill can be decided. Toilet windows can be planned at proper heights. To prevent the street noise, a barrier like compound or boundary wall can be planned of proper height. To cut down the street noise and dust, planting trees around the house will be effective. There are many ways to ensure external privacy in the building.

(b) Internal privacy: this aspect helps prevent direct view from other room or passage. Proper placement of door openings, placement or door shutter or other screening like furniture arrangements helps in maintaining internal privacy.

2.1.6 Elegance

Elegance is the aspect in the building which can be planned only by elevation like the front façade of the building to make it appealing and pleasing. To make elevation elegant and attractive, principles of regularity, symmetry is planned with decorative plaster, cladding or paint on the external walls etc. While planning the building, the elevation for elegance is one of the important principles of planning. Length and height ratio also plays an important role in enhancing the beauty of the elevation.

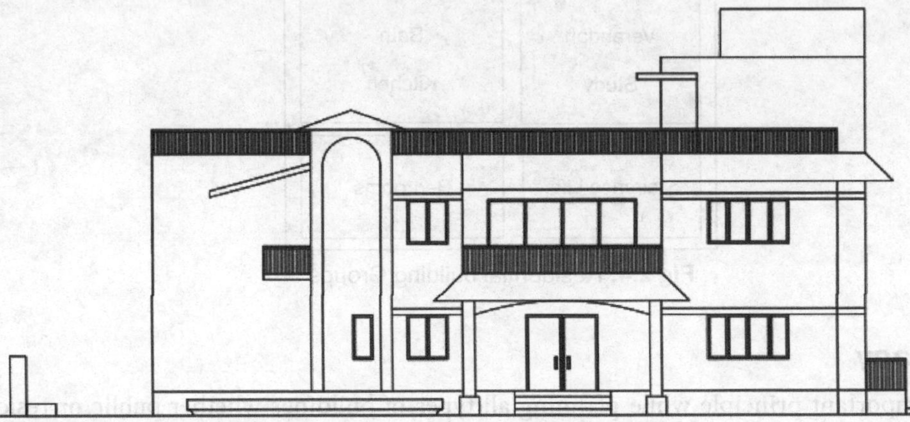

Fig 2.6: Elegance in Elevation

2.1.7 Flexibility

In planning, if a space is planned for one function, and it can be converted to different function then there is flexibility built in. In residential building planning flexibility aspect is not necessary however for some gatherings, one can convert small rooms into bigger by planning folding doors or partition as required.

2.1.8 Circulation

Circulation means movement from one place to another it can be horizontal or vertical.

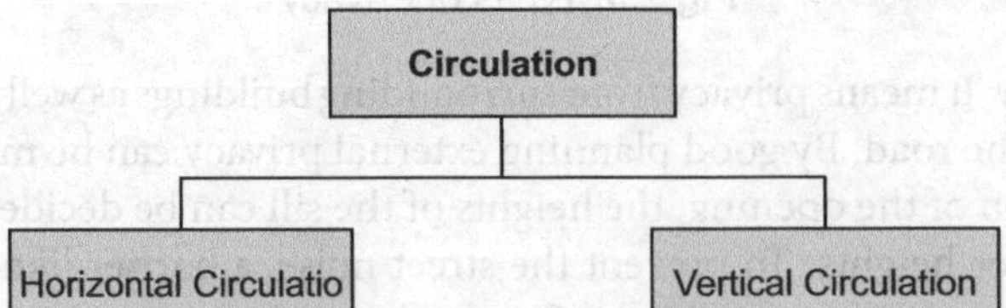

Fig 2.7: Types of Circulation

(a) ***Horizontal circulation***- It refers to movement of person from one room to another with the use of doors, corridors, passages, etc. the ratio between circulation space to utilization space should be around 1:4 for economic planning for hostels, educational institutes, offices factories it can be 1:3. For example, corridors can be properly designed to go from one classroom to another.

(b) ***Vertical circulation***- In multistory building, vertical circulation has to be properly planned. It refers to movement of person from one floor to another with the use staircase, lift, ramps, escalators, etc. About 8-10 % of the floor area is adequate to plan vertical circulation. For example, generally staircases and lifts are provided in building for vertical circulation.

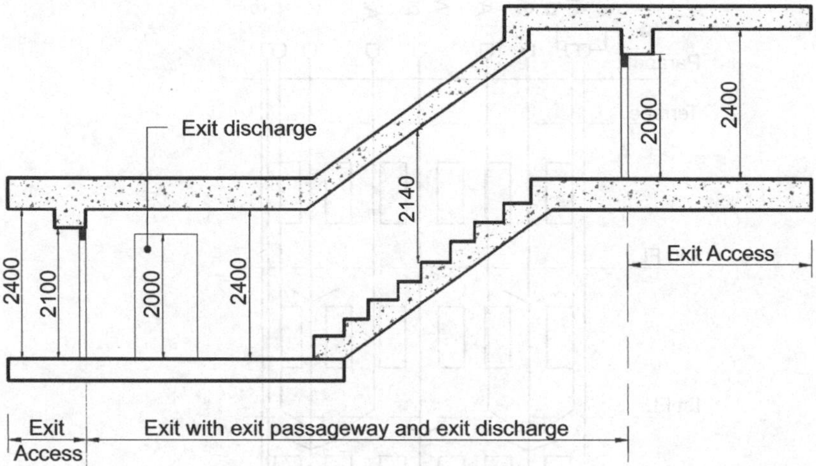

Fig 2.8: *Vertical Circulation*

2.1.9 Furniture Requirements

Furniture planning is one of the important requirements while planning the space. The room size can be decided on number of users, furniture requirement, equipment, arrangement of furniture and space for circulation needs to be planned efficiently.

The furniture requirement for a dining room and a bedroom depends upon the function of that room for residential buildings. In case of public building the space should be sufficient to meet the needs of the requirement. For example, the classrooms for a particular number of students to be designed. So accordingly, furniture arrangement is decided.

2.1.10 Sanitation

Sanitation means to take care of light, ventilation, air conditioning, thermal comfort, facilities for general cleaning and sanitary conveniences. Sanitary conveniences include water closets (WC), bathrooms, lavatories, latrines, urinals etc. Good ventilation in a building is an important factor to consider while planning any building. In case of public buildings like schools, hospitals, factories etc. maximum natural ventilation is to be aimed while planning.

Each room must be planned to provide 10-15% of the floor area for openings. The ratio of window area to the floor area needs to be planned according to the climate of the region.

Table 2.2: Window to floor area

Sr. No	Climatic Region	Window to floor area
1	Hot and Humid	Window area $1/5^{th}$ of floor area of the room
2	Hot and Dry	Window area $1/6^{th}$ of floor area of the room
3	Wet and Heavy rainfall	Window area $1/7^{th}$ of floor area of the room

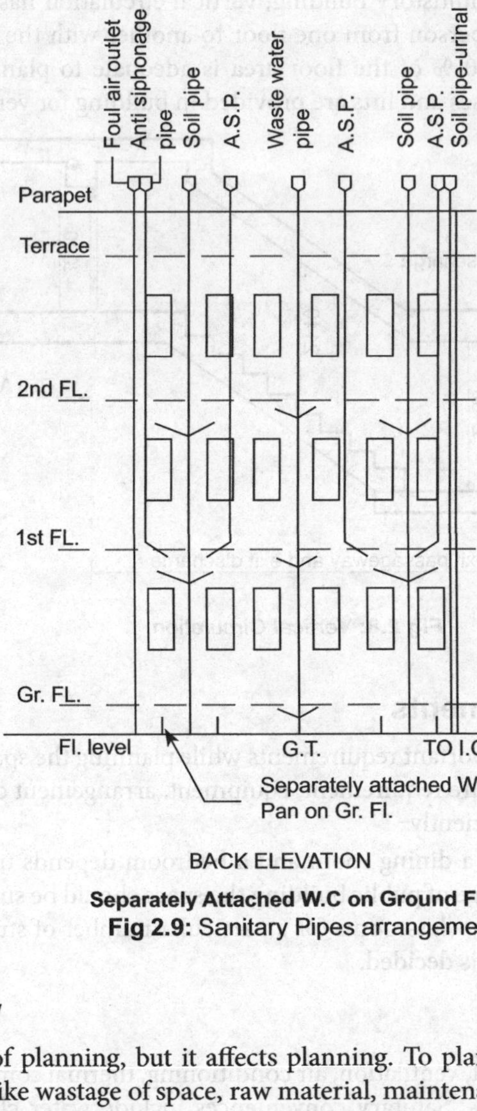

Separately Attached W.C on Ground Floor
Fig 2.9: Sanitary Pipes arrangement

2.1.11 Economy

This is not a principle of planning, but it affects planning. To plan any building economically means avoid all types of waste like wastage of space, raw material, maintenance. Judicious use of space will lead to efficient and economic planning. Building should be well planned and after construction it should require less maintenance for an economic planning.

2.2 SPACE REQUIREMENTS AND NORMS FOR MINIMUM DIMENSION OF DIFFERENT UNITS IN THE RESIDENTIAL AND PUBLIC BUILDINGS AS PER IS 962.

Space requirements gives an idea about the distribution of spaces for different functional needs. A planned growth in a development gives healthy lifestyle to the citizens, Protects life, health, and welfare of occupants. Any structure used for humans or not, constructed with any material for any purpose and includes foundation, plinth, walls, floors, roofs, terrace, verandas, balconies, cornices, projections, plumbing, and other services are all part of a building. Public building means where groups of people gather for any purpose, it can be amusement, recreation, social, religious, patriotic, civil, travel and similar purposes.

To construct residential building certain specific spaces must be allocated and similarly for public building some norms must be followed. The requirement of space for residential and public buildings follows traditional norms and these norms change with time too.

These are general guidelines which suggest minimum requirements of different spaces of the buildings.

For Residential Buildings

2.2.1 Staircase

Riser and Tread in Residential Building:

Riser : 175-185 mm

Tread : 250-270 mm

These are keeping in mind average comfort of feet rise and using the stairs.

2.2.2 For Different Elements

Table 2.3: Standard heights

Sr. No.	Elements	Standard heights
1	Plinth Height	0.60 – 0.75m
2	Windowsill	0.60m
3	Windowsill for WC and Bath	1.20m
4	Window height	1.5m
5	Door Height	2.1m
6	Floor to ceiling height	3-3.5m
7	Head room for staircase	3.00m
8	Slab Thickness	100-120mm
9	Kitchen platform	0.80m
10	Riser of the steps and staircase	150 -180mm

Table 2.4: Area, Minimum width, Ceiling heights

Sr. No	Room	Minimum Area in m^2	Minimum width	Minimum Height
1	Habitable room	9.5m^2 (for single room)	2.4m	2.75 m
2	Kitchen	5.5 m^2	1.8 m	2.75 m
3	Bathrooms, Water Closets, combined Bath & WC (Measured from the surface of the floor to the lowest point of the ceiling)	1.1 m^2	0.9 m	2.2 m

Sr. No	Room	Minimum Area in m²	Minimum width	Minimum Height
4	Head Room		1.8 m	2.2m
5	Air-conditioned habitable room.		2.4m	4.2m
6	Assembly halls, residential hotels of all types, institutional, educational, industrial, hazardous or storage occupancies, departmental stores, malls, entrance halls and lobbies to departmental stores and Data Centre		3.6m	6.0m
7	IT buildings, office buildings		2.75m	4.2m
8	Exhibition cum Convention Centre, Sport facility requiring more height		4.2m	8.8m
9	Sound Recording/Film Studio, Warehouse		4.2m	12m
10	Pitched roof – (a) Any habitable room		2.75 (average with 2.4 m at the lowest point)	4.2 (Average with 2.8 m at the lowest point).

For Public Buildings

Public buildings are designed for the many people. Common arrangements need to be done like

1. Staircase
2. Sanitary block
3. Circulation
4. Lifts
5. Entrance or reception
6. Parking space, cycle stands, garages
7. Watchman cabin

2.2.3 Staircase

Riser and Tread in Public Building:

Riser : 150-170 mm

Tread : 270-300 mm

2.2.4 Sanitary Block

It includes bathroom, WC, wash basin and urinals. According to the requirements the number of the units may vary.

Table 2.5: Bathroom, WC, wash basin and urinals

Bath room	1.20 x 2.10m 1.50 x 2.40m 1.90 x 2.70m
WC	0.90 x 1.20m 1.0m – 1.20m
Urinals	0.90 – 0.75m
Passage width	0.90 – 1.50m

2.2.5 Circulation

Various room are joint together horizontally and vertically both. for horizontal movement passages, corridors, and verandah .

Table 2.6: Movement Passages

Passage - corridor	1.0m - 2.5m
Verandah	1.80m – 3.50m
Stairs	Width 1.20m(minimum) Riser 150- 170mm Tread 300-325mm Landing 1-1.80m wide Headroom 1.80m minimum
Floor height	2.75 – 6.00m

2.2.6 Lifts

For vertical circulation lifts are quick and efficient. There are different types of lifts are available. According to the character of the building, frequency of service required lifts are to be selected

Lifts are of many types

- Lifts used in office buildings, hotels, and high-class flats to carry 10-20 persons
- Lifts used in shops and departmental stores
- Lifts for hospitals passenger and stretcher lifts
- Lifts for fire
- Lifts to carry goods (dumb waiter)

Lifts used in low and medium class flats for 5 persons

- Car lift speed – 0.5 m/s
- Door for lift – sliding
- Landing doors – closed
- Method of control – automatic push button

Lifts used in office buildings, hotels and high class flats to carry 10-20 persons

Table 2.7: Lift speed

No of floors	Lift car speed in m/s
Up to 2	0.5
3-4	0.5 – 0.75
5-6	0.75-1.5
7-9	1.5 - 2
10-12	2-2.5
Over 12	2.5 - 3

For car speeds exceeding 1.75m/s, gearless machines should be provided, car door – automatic type

Method of control – car switch control, automatic push button

Lifts used in shops and departmental stores

Car speed 2 – 2.5m/s

Door for lift – automatic opening

Method of control – car switch control

Lifts for hospitals passenger and stretcher lifts

Passenger lift for goods and persons

Car speed – 0.25 – 0.5m/s

Gates for the car – collapsible

Control method – depends upon the local conditions

Hospital bed lift or stretcher lift

Car speed – 0.25 – 0.5m/s

Car door – solid sliding

To facilitate the wheelchair and stretcher in and out of the lift, an automatic car levelling device is also included.

2.2.7 Entrance or Reception

Some space is required at the entrance in every public building.

The area of this entrance varies with the number of people entering at one time. The width of the entrance should not be less than 2.75m.

General size of the entrance or reception are 3 x 6m, 3.5 x 7.0m, 3.75 x 8.0m, 4 x 5.0m, 4.5 x 6.0m, 6x8.0m

2.2.8 Parking Spaces, Cycle Stand and Garages

Open parking is required in any type of public building. Area required for parking will depend upon nature of the building and number of people visiting the building.

For every public building provision of cycle stand and garage is an essential requirement.

Table 2.8: Area required for parking for different vehicles

Vehicles	Area Required/ vehicle
Car	$20m^2$
Scooter and Motorcycles	$3m^2$
Cycles	$1.2m^2$

Driving aisles (passage between two blocks) for cars: 1-2m wide

Garages: 3m x 4m, 3.3 m x 4.2 m, 4.0m x 4.5 m

Watchman's Room: this should be located near the entrance gate of minimum 2 x 2m.

2.3 RULES AND BYELAWS FOR SANCTIONING AUTHORITIES FOR CONSTRUCTION WORK

While building the house, there are laws or regulation decided by the city authorities for that city which are to be followed. If rules and byelaws are not followed, then the houses may affect the interest of others in respect to health and convenience resulting in haphazard situation. In absence of laws and regulations, the landlord will want to spend minimum amount to get maximum benefit while the builder will want to take maximum possible advantage and not provide adequately for amenities like drainage line, water supply line, roads etc putting excessive load and creating unhealthy environment. Thus, there is a restriction to regulate the power for the builder or owner and follow certain rules.

Building bye laws are the rules laid down for construction and planning of different types of buildings by municipal corporation, town planning authorities or revenue authorities. These rules can be different for different places, like Mumbai byelaws are different from some district or taluka having municipality or village having gram panchayat. After scrutinizing the plans regarding the building byelaws, these authorities sanction the building plan.

2.3.1 Necessity of Building Bye Laws

- To reduce the haphazard growth of the city and town, to control land development and to check the un-authorized constructions.
- To define the rules for new structures to be built.
- To specify the type of materials for construction to be used.
- To provide for widening of roads and other development for future.
- By providing hygienic environment and also reduce the pollution by restricting population density in the town.
- To provide for every citizen to get basic facilities like water supply, drainage, electricity, proper light and ventilation etc.
- To provide for open spaces, and safety against fire, noise, smoke etc.

Scan Me

2.3.2 Open Space Requirements

Around the building the open spaces is required for light, ventilation, future expansion and also to approach other property.

Setbacks or open spaces in front, rear and side of the building depends upon the height of the building and can be calculated by

W = Width of open space around the building in m = 3+(h/10)/3

Where h is the height of the building in m < 25m

Open space should not be less than 1.8m for rear side of the building of height less than 10m.

Setback Distance: It is measured from centre line of the road till the plinth of the building.

It is demarcated for future expansion of the road, road widening, parking of vehicles, circulation, or air.

2.3.3 For Public Buildings, Commercial Buildings the Setback is More Than Residential Buildings

Table 2.9: Setback Distance

Type of Roads	Minimum setback distance		Ratio of column 3 to column 2
	Residential building	Industrial building	
Village Road	9m	15m	1.67
Major district Road	15m	24m	1.60
National or state Highway	30m	45m	1.50

2.3.4 Height of the Building

maximum height of the building depends upon the width of the road or street which is in the front. Minimum width of the rear space and proximity to the airport.

Table 2.10: Height of the building

Width of the street	Maximum height of the building
<8m	1.5 times width of the street
8m to 12m	12m
>12m	<24m

2.3.5 Bathroom and Water Closets (WC)

- No bathrooms or water closet shall be situated directly over any room other than another water closet, washing place, bathroom or terrace unless the said floor is made impervious with adequate water-proofing treatment. However, in no case shall a water closet or bathroom be provided over a kitchen.
- Every bathroom or water closet shall have the platform or seat or flooring made of water-tight non-absorbent material.
- It shall be enclosed by walls or partitions and the surface of every such wall or partition shall be finished with a smooth impervious material to a height of not less than 1m above the floor of such a room.
- It shall be provided with an impervious floor covering sloping towards the drain with a suitable grade and not towards a verandah or any other room.

- Every water closet and/or a set of urinals shall have a flushing cistern of adequate capacity attached to it.
- All the sewerage outlets shall be connected to the municipal sewerage system.

2.3.6 Loft

- Location: Lofts may be provided over kitchens, habitable rooms, bathrooms, water closets and corridors within a tenement in residential buildings, in shops and in industrial buildings.
- Height: The height of the loft shall not be more than 1.5 m. If it exceeds 1.5 m, it shall be counted towards FSI.
- The lofts in non-residential buildings shall be located at least 2 M away from the entrance.

2.3.7 Garage

- Size-The size of a private garage shall not be less than 2.5 m x 5.5 m or 2.3 m x 4.5m.
- Location-If not within the building the garage may be located at its side or rear, but at least 7.5 m away from any access road.

Table 2.11: Minimum width of common staircase and corridors for various residential and Public Buildings

Sr.No	Type of occupancy	Minimum width of corridors /stairways in building less than 70m	Minimum width of corridors /stairways in building more than 70m
1	Residential buildings		
	(a) General	1.50	2.00
	(b) Row housing (2 storied) or internal staircase in duplex flat	0.9	
	(c) Hotels		
2	Educational buildings		
	(a) Up to 32 m high	1.50	
	(a) over 32 m high	2.0	
3	Institutional Buildings	1.50	2.0m
4	Assembly Buildings	2.0m	2.0m
5	Mercantile storage, hazardous buildings	1.50	2.0m
6	Industrial Building	2.00m	

2.3.8 Ramps

Ramps for pedestrians-

A ramp should be provided as per the staircase norms. A ramp in the hospital should be 2.25m wide and the slope not less than 1in8 for vehicular traffic and 1 in 12 for pedestrian with non-slippery finish.

(a) Handrail- A handrail shall be provided on both the sides of the ramp with minimum height of 0.90 m above the surface of the ramp.

(b) For every 9 m of ramp length, a landing of 1.5 m shall be provided

(c) Landing of minimum 1.5 m length shall be provided at every entry & exit of ramp.

Ramps for Vehicles

Table 2.12: Width of Ramps for vehicles

Sr. No.	Type of Vehicle	Clear width of Ramp in meters (m)	
		One Way Ramp without central divider	Two Way Ramp without central divider
1	Light Motor Vehicle (LMV)	3.00	6.00
2	Medium Commercial Vehicle (MCV)	3.50	7.00
3	Heavy Motor Vehicle (HMV)	4.50	9.00

2.3.9 Parapet

Parapet walls and handrails provided on the edges of the roof terrace, balcony, etc. shall not be less than 1.50m from the finished floor level.

2.3.10 Special Design Requirement for Educational Building

In addition to the classrooms and other areas, every educational building shall be provided with-

(a) a tiffin room with a minimum area of 18 sq. m for every 800 students or part thereof;

(b) a separate tiffin room for teachers where strength of students exceeds 1000.

(c) a room/space with drinking water facilities for every 300 students or part thereof on each of the floors.

2.3.11 Special Design Requirement for Building for Medical Use

For Hospital, Maternity Homes and Health Centre, Sanatorium, Multi-Specialty Hospitals: minimum area of general wards shall be 40sq.m with no side less than 5.5 m;

2.3.12 Special Regulations for Differently Abled Persons

Differently abled people are if they come under any such category seven categories of disabilities have been identified viz., blindness, low vision, leprosy cured, hearing impairment, locomotor disability, mental retardation and mental illness. The provisions made for Differently Abled People are also applicable to elderly, pregnant women and mothers with small children on prams.

Scope:

These bye laws are applicable to all existing and proposed public buildings & facilities having BUA more than 2000 sq. m.

Site Development:

Level of the roads, access paths and parking areas shall be described in the plan along with specification of the materials.

Access path/walkway:

Access path from plot entry and surface parking to building entrance shall be of minimum 1.5 m width having even surface without any slope. Slope if any shall not have gradient greater than 1 in 20. Finishes shall have a non-skid surface with a texture traversable by a wheelchair. Where transfer must be made from a vehicular surface to a pedestrian surface, the driveway and the pavement or footway surfaces shall be blended to a common level or ramp.

Table 2.13: Gradient and width of Ramp

Level Difference	Minimum gradient	Ramp Width	Comments
≥ 0.15 m < 0.30 m	1:12	1.2m	
≥ 0.30 m < 0.75m	1:12	1.5m	Landing every 5m of ramp run
≥ 0.75 m < 3.0 m	1:15	1.8m	Landing every 9m of ramp run

Parking:

For parking of vehicles by Differently abled individuals, the following provisions shall be made:

(a) Minimum two surface parking spaces for the first 100 parking spaces and one parking space for next 200 parking spaces or part thereof shall be provided near entrance for the Differently abled persons with maximum travel distance of 30 m from building entrance.

(b) The width of parking bay shall be minimum 3.6 m.

(c) The information stating that the space is reserved for wheel chair users shall be conspicuously displayed.

(d) Indication/directional signage along driveway showing the way leading to the parking spaces reserved for Differently abled persons should be provided.

Building Requirements:

The specified facilities for the buildings for Differently abled persons shall be as follows:

1. Approach to plinth level.
2. Corridor connecting the entrance/exit for the Differently abled persons.
3. Stairways.
4. Lift.
5. Toilet.
6. Drinking Water

Stair Ways:

One of the stairways near the entrance/exit for the Differently abled persons shall have the following provisions:

(a) The minimum width shall be1.5 m.

(b) Height of the riser shall not be more than 0.15 m and width of the tread 0.3 m. The steps shall not have abrupt (square) nosing.

(c) Maximum number of risers on a flight shall be limited to 12.

(d) Handrails shall be provided on both sides and shall extend 0.3 m on both sides and shall extend 0.3 m on the top and bottom of each flight of steps.

(e) The minimum clear headroom in pedestrian area such as walkways, halls, corridors, passageways, or aisles shall be 2.2 m.

Lifts:

Wherever lift is required as per byelaws, provision of at least one lift shall be made for the wheel chair user with the following cage dimensions:

Clear internal depth: 1.2 m
Clear internal width: 1.4 m
Entrance door width: 0.9 m

(a) A handrail not less than 0.6 m long at 1.0 m above floor level shall be fixed adjacent to the control panel. Also, switch control /call button shall be at an operating height of 0.9 m to 1.2 m.
(b) The lift lobby shall be of an inside measurement of 1.8 m x 1.8 m or more.
(c) The time of an automatically closing door should be minimum 5 second and the closing speed should not exceed 0.25 m/sec.

The gap between the lift floor and floor landing shall not be more than 12 mm.

Toilets:

One special W.C. in a set of toilets with non-skid surface shall be provided for the use by the Differently abled persons with essential provision of wash basin near the entrance: -

(a) The minimum size shall be 1.5 m x 1.75 m
(b) Minimum clear opening of the door shall be 0.9 m and the door shall swing out.
(c) Suitable arrangement of vertical/horizontal handrails with 50 mm clearance from wall shall be made in the toilet.
(d) The WC seat shall be 0.5 m from the floor. The flushing control shall be mounted on the wide side of the cubicle at a height from 0.6 m to 1.05 m above the finished floor level.

Special requirements at public spaces like auditorium and theatre

Table 2.14: Following procedure should be adopted for getting approval for building construction from competent authority.

Provision & Location	a. Guideline: All theatre, concert hall, places of recreation and auditoriums etc areas should accommodate Differently abled persons using various mobility aids.
	b. At least minimum of two wheelchair spaces shall be provided for every 200 seats or less. Additional one wheelchair space shall be provided for every subsequent addition of 200 seats.
Dimensions	c. While planning the internal spaces Wheelchair spaces should be plan with any seating plan. And distributes in the different pricing rows to not obstruct view of the person behind.
	To sit with the able-bodied companion, the wheelchair space, 0.9 m x 1.2 m should be plan.

1. **Notice:** Every person who intends to carry out development and erect, re-erect or make alterations in any place in a building or demolish any building, shall give notice in writing to the Planning Authority of his said intention in the prescribed form and such notice shall be accompanied by the payment receipt of required scrutiny fee and any other fee/ charges prescribed by the Planning Authority from time to time and the plans and Statements in sufficient copies. The plans will be sent in the printed form. One set of plans shall be retained in the office of the Planning Authority for record after the issue of permit or refusal. For the sake of scrutiny, the plans may be submitted in the form of soft copy as specified by the Planning Authority from time to time.
2. **Copies of Plans and Statements:** Minimum four copies of plans and statements shall be made available along with the notice. In case of building schemes, where clearance is required from

other agencies like Fire Services, number of copies of plans required shall be as decided by the Chief Officer.

3. **Information Accompanying Notice:** The notice shall be accompanied by the key (location plan), site plan, sub-division layout plan, building plan, services plans, specifications and Certificate of supervision and ownership title as prescribed in Regulation
4. **Ownership title and area:** for development permission and commencement certificate shall be accompanied by the following documents for verifying the ownership and area etc. of the land.
   (a) Attested copy of original registered sale / lease - deed / power of attorney / enabling Ownership document wherever applicable.
   (b) V.F.No.7/12 extracts or property register card of a date not more than six months prior to the date of submission and a certified copy of the Measurement Plan of the property under Development proposal.
   (c) Statement of area of the holding by triangulation method from the qualified licensed technical personnel or architect with an affidavit from the owner with regard to the area in the form prescribed by the Chief Officer.
   (d) Any other document prescribed by the Chief Officer.
   (e) Wherever third-party interest is created by way of agreement to sale or mortgage etc. The Registered consent of such interested persons shall be submitted with the application.
   (f) A certified copy of approved sub-division / amalgamation / layout of land from the Concerned authority.
5. **Key Plan or Location Plan:** A key plan drawn to a scale of not less than 1:10,000 shall be Submitted as a part of building plan / development proposal along with the application for a Building permit and commencement certificate; showing the boundary and location of the site with respect to neighborhood landmarks or with respect to the area within the radius of 200 from the site, whichever is more.
6. **Site Plan:** For building permission, the site plan drawn to scale of 1:500 or more shall be submitted with an application or decided by the Chief Officer. This plan shall be based on the measurement plan duly authenticated by the appropriate officer of the Department of Land Records.
7. **Sub - Division/ Layout Plan:** For the development of land, the notice with the subdivision/ layout plan drawn to a scale of not less than 1:500 is to be submitted, and for layout having areas of 4.0 ha. and above, the plan shall be drawn to a scale of not less than 1:1000.
8. **Building Plan:** The plans of the buildings its elevations and sections to be drawn to a scale of 1:100 and sent with the notice.
9. **Service Plan:** For big layouts plans not less than 1:1000 and the scale of not less than 1:100 as required by the authority, the elevations and sections of private water supply, sewage disposal system and details of building services needs to be submitted.
10. **Specifications:** For the proposed constructions, general specifications are required for type and grade of materials to be used.
11. **Supervision:** A certificate of supervision in the prescribed form by a licensed Architect/ Engineer/ Structural Engineer is to be given with the notice.

12. **Building Permit Fee:** An attested copy of Receipt of payment of Building Permit Application Fee is attached to the notice.
13. **Security Deposit Fee:** A security fee shall be charged as per the rates specified by the Chief Officer for faithful compliance of regulations and the directions given in the sanctioned plan.
14. **No Objection Certificate:** No objection certificates are required from Civil Aviation Authority, Railways, Directorate of Industries, Maharashtra Pollution Control Board, District Magistrate, Inspectorate of Boilers and Smoke Nuisance, Defence Authorities, Maharashtra Coastal Zone Management Authority, Archaeological Department etc. for any development / construction of buildings. These clearances from the authorities are required and attached to the application.
15. **Signing the Plan:** All the plans shall be duly signed by the owner/ co-owner and the Architect or Licensed Engineer / Structural Engineer / Supervisor who shall indicate his name, Address and license number allotted by the Chief Officer.

    Documents and drawing required for obtaining sanction from competent authority.

    - ☐ Plan
    - ☐ Site plan
    - ☐ Elevation
    - ☐ Section
    - ☐ Detailed estimate
    - ☐ Details of electricity, sewerage arrangement
    - ☐ Ownership right.

Site plan, location Plan and area statement includes

- To know the orientation of the building.
- All floor plans including terrace and car park plan with North indication.
- Elevation architectural elevation of the building.
- Section for internal details from foundation to terrace of the building also total height of the building.
- Schedule of openings to give details of the doors and windows sizes and types.
- Schedule for type of construction.
- Schedule for foundation work.

  It should also show

  - ☐ Total area, floor area
  - ☐ Details of Doors, window, ventilation
  - ☐ Foundation details
  - ☐ Character such as residential and commercial etc.

The department checks each and every detail on the plan and points out for compliance or any other deficiencies. After compliances with the requirements of the building byelaws are fulfilled, they get the approval and licence is issued with the following condition:

- There should be no deviation, alteration, addition, or any change from the sanctioned plan.
- The copy of sanctioned plan must be submitted to the appropriate authority.
- The completion certificate must be obtained within one month of the completion of building construction for the assessment of house tax.

All building byelaws are always prospective in nature and are applicable only from the date of issue.

2.4 PLOT AREA BUILT UP AREA, SUPER BUILT-UP AREA, PLINTH AREA, CARPET AREA, FLOOR AREA AND FAR (FLOOR AREA RATIO)

2.4.1 Plot Area

A piece of land demarcated by definite boundaries. Any shape or size of land is called plot and the area of this land is called plot area. Plots are sold either by competent authority or development authority or by owner of land. After purchasing, to prevent from further encroachment by any other antisocial elements, it should be well marked and fenced. These days it is good to ensure that the digital maps are available with authorities.

Scan Me

2.4.2 Built Up Area

The area covered by all floors of the building is called as built-up area. excluding balconies, staircases, etc. It is floor area of all rooms plus wall thickness.

2.4.3 Super Built Up Area

super built-up area is the sum of built up area. And certain percentage of floor area marked for staircase, Lift, parking and open balconies as decided by the local development authorities.

2.4.4 Carpet Area

Carpet area is the usable rooms at any floor. Actually, it is the area where carpet can be laid. It excludes area of sanitary accommodation, verandah, corridor and passages, kitchen and pantries, stores, entrance and porches, staircase and mumty, shafts for lifts, barsaties, garages, canteens, air conditioning ducts and plant room.

2.4.5 Floor Area

Floor area is calculated by deducting area of walls from plinth area. This is the usable covered area of the building at every floor level.

Floor Area = plinth area – walls

Areas to be included

- Door and other openings
- Plaster on the wall if exceeds 300sqm.
- Internal columns.

Area not to be included

- Plaster on the wall if greater than 300sqm.
- Projections beyond the face of the wall.

2.4.6 Plinth Area

Plinth area is the area of the building with wall thickness at plinth level. It is the raised platform of the structure on which building is going to be constructed. This is built up covered area measured at the ground floor level.

Following areas shall be calculated

- When building is constructed in frame structure areas of walls at floor level excluding plinth offsets, projections beyond cladding.
- For sanitary installation, air conditioning ducts, lifts act as the Internal shaft if area is less than $2m^2$ in area.
- Porches and cantilevers
- Area of Mumty / Barsati at terrace

Following areas shall not be included

- Internal shaft if area is more than $2m^2$ in area.
- Uncovered balconies
- Architectural bands, cornices
- Vertical fenestrations

2.4.7 FAR

Floor Area Ratio is a ratio of total built up area to the area of plot. It is also called as floor space index (F.S.I). F.S.I. means floor space index.it varies from city to city and also area to area. To regulate population density and overcrowding in dwelling units it restricts the floor area of the building in relation to its plot area. If FSI is 1, then total permissible area on all the floors of the building is equal to the area of the plot.

FSI = Total built up area on all the floors / Total plot area

2.5 LINE PLANS OF A RESIDENTIAL BUILDING OF MINIMUM THREE ROOMS INCLUDING WATER CLOSET (WC), BATH AND STAIRCASE AS PER PRINCIPLES OF PLANNING

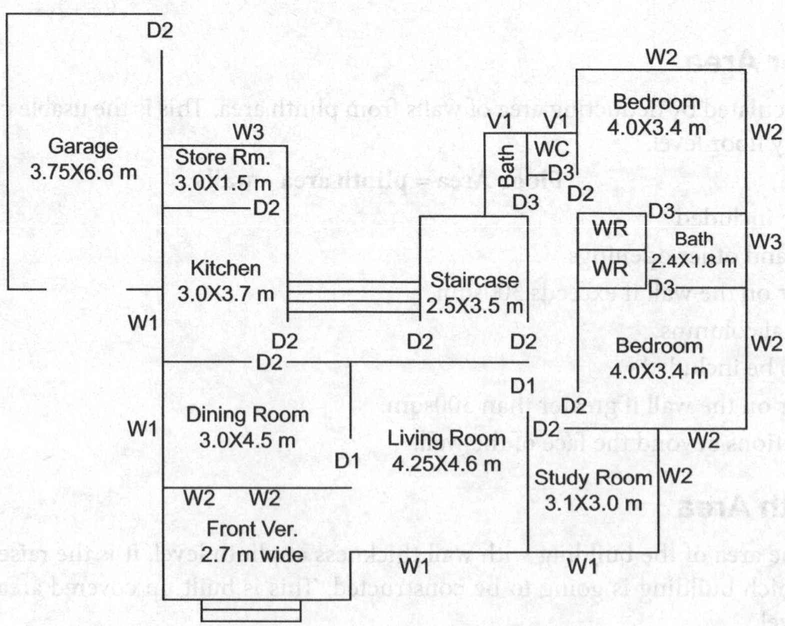

Fig 2.10: Line plan of a Residential Building

D1	1M X 2.1M
D2	0.8M X 2.1M
D3	0.75M X 2.1M
W1	2M X 1.5M
W2	1.2M X 1.5M
V1	0.6M X 1.5M

Scan Me

2.6 LINE PLANS OF A PUBLIC BUILDING SCHOOL BUILDING, PRIMARY HEALTH CENTRE, RESTAURANT, BANK, POST OFFICE, HOSTEL, FUNCTION HALL AND LIBRARY

2.6.1 Requirements of School Building Can be for Primary and Secondary School

Sr. No	Spaces	Area
1	Entrance or Reception	9-12m^2
2	Office and Administrative Block	12-15m^2
3	Classrooms Students below 10years Students above 10years General Sizes	Area 1-1.2 m^2/student Area 1.2-1.5 m^2/student 4.5m-6 m 5.5m-6.5 m 6.0m-7.2 m 6.0m-7.8 m 7.0m-10 m
5	Teachers Room	3.0mx 4.5m
6	Drawing Hall	3.0 – 4.0 m^2/student
7	Assembly Hall	0.5 – 0.6 m^2/student
8	Circulation	1m-2m
9	Library	90-100m^2 for 1500 students
10	Science Lab	3.0-4.0m^2 /student
11	Parking space and cycle stand	
12	sanitary units	

Description	Male	Female
WC	1 for 40 students	1 for 25 students
Urinals	1 for 20 students	
Wash Basin	1 for 40 students	1 for 50 students
Water Taps	1 for 40 students	1 for 50 students

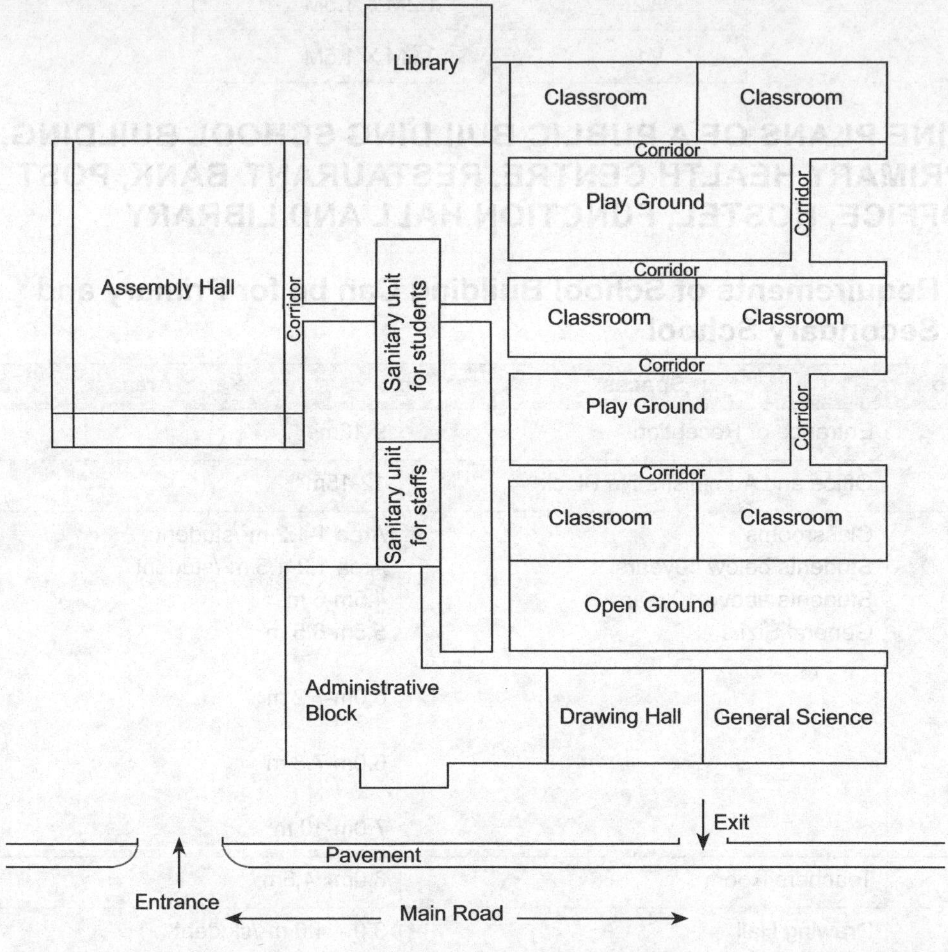

Fig 2.11: Line Plan of School Building can be for primary and secondary school

2.6.2 Requirements of Primary Health Centre

Sr. No	Spaces	Area
1	Entrance and waiting space	3m x 3.6m
2	Doctor's Room	3m x 4m
3	Examination Room	3m x 4m

Sr. No	Spaces	Area
4	Operation Theatre	4m x 5.5m
5	Circulation Space	2m x 3m
6	Wards: Maternity, General	6 x 8m²/bed
7	Medical Store	3.5m x 4.75m
8	Office	12m²-15m²
9	Laboratory	12m²-15m²
10	Residence: Doctor Servants	 60-90 m²/ head 40-60 m²/head

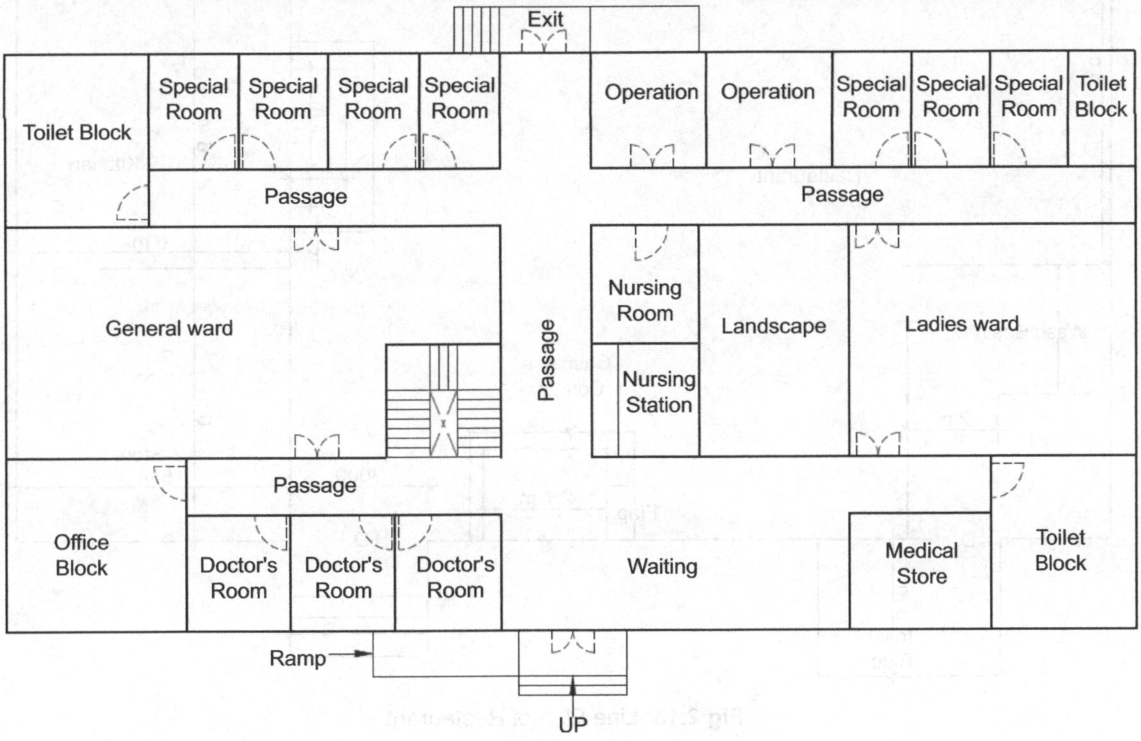

Fig 2.12: Line Plan of Primary Health Centre

2.6.3 Requirements of Restaurant

Sr. No	Spaces	Area
1	Entrance and Reception	
2	Waiting Hall	
3	Restaurant /Eating Space	150-180 m²
4	Kitchen	25-30 m²

Sr. No	Spaces	Area
5	Store	25-30 m^2
6	Pantry	6 m^2
7	Circulation	1.5-2.0 m
8	Sanitary Block Male and Female	5-8 m^2
9	Parking Space	

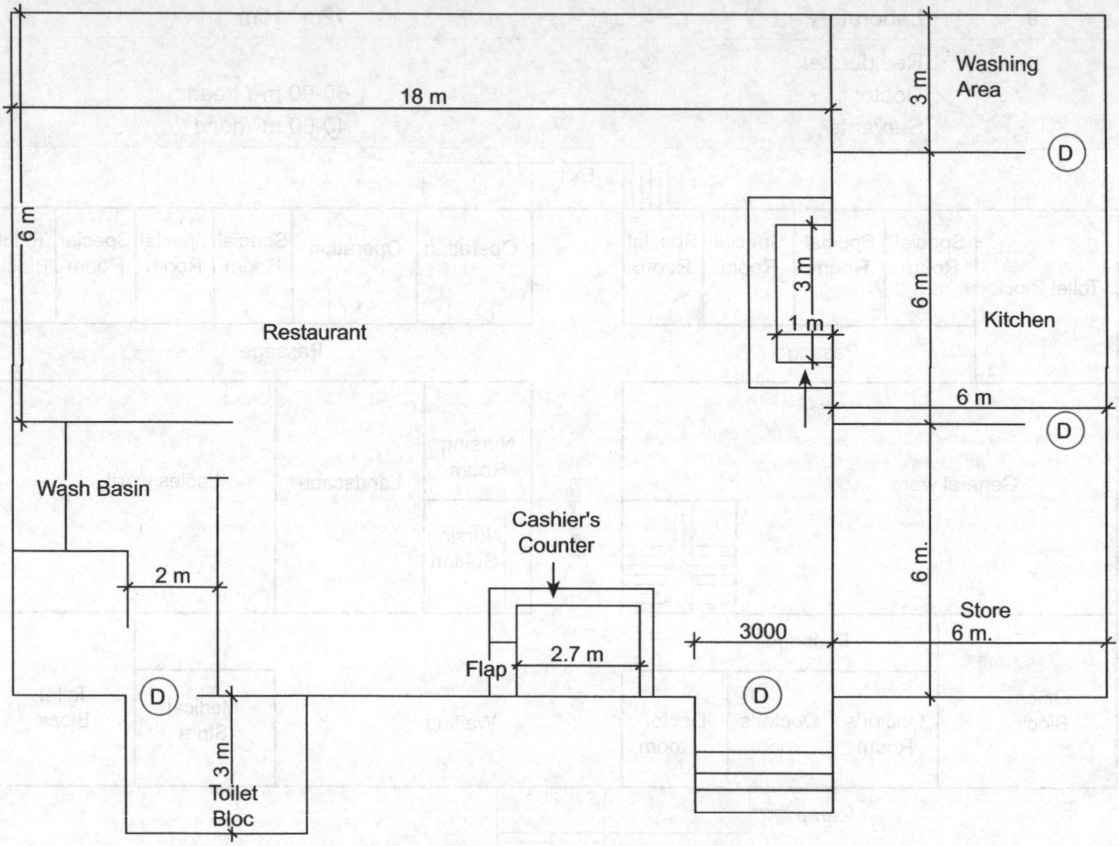

Fig 2.13: Line Plan of Restaurant

2.6.4 Requirements of Bank

Sr. No	Spaces	Area
1	Entrance and moving space	2m Wide (Minimum)
2	Counters Height width	1.6m-1.8m 0.4m-0.8m
3	Working space behind the counters	3m wide
4	Strong room	9-12m^2

Sr. No	Spaces	Area
5	Safe deposit lockers/store room	4m x 3m
6	Managers Room	3.5m x 3.5m
7	Cashier's cabin	1.5m x 1.5m
8	Toilets For Male Female	1.5m x 3.0m

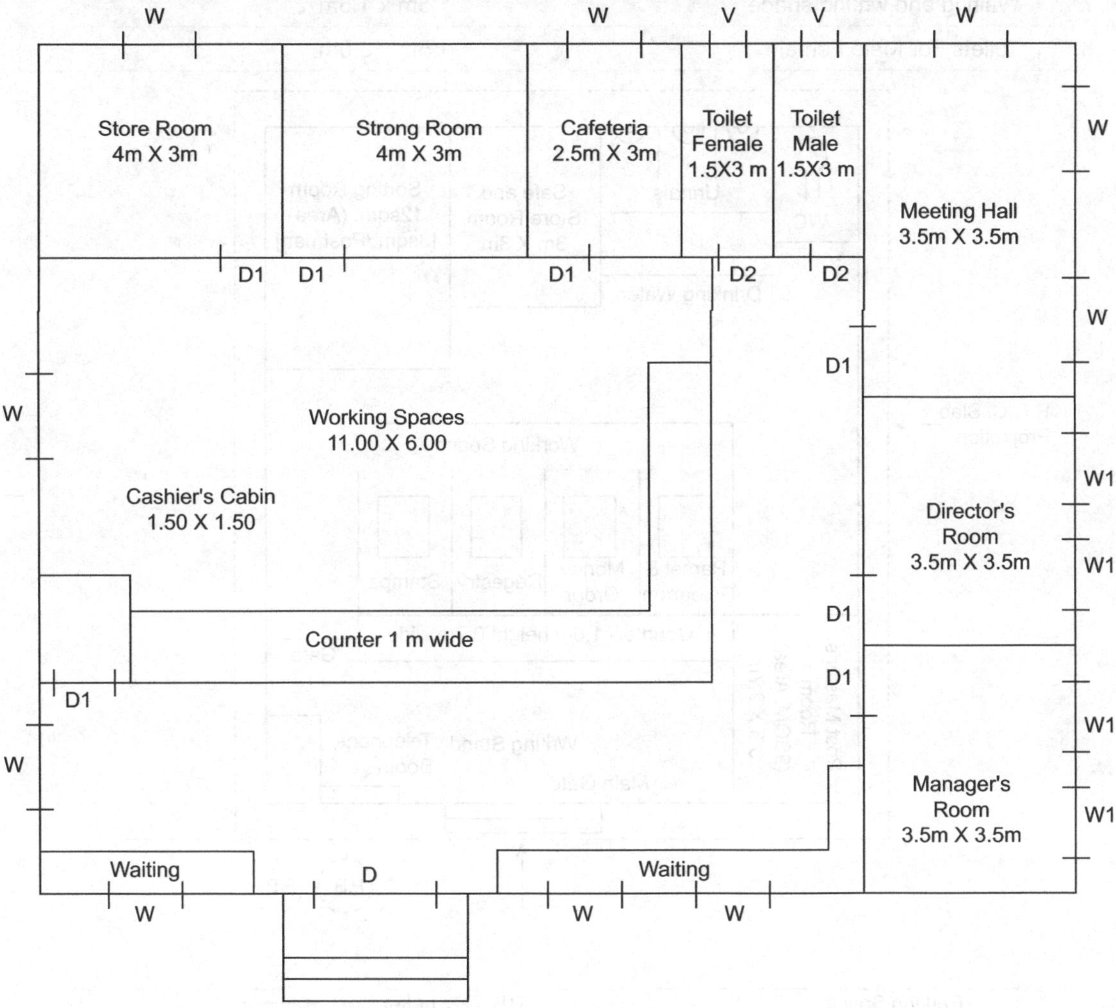

Fig 2.14: Line Plan of Bank

2.6.5 Requirements of Post Office

Sr. No	Spaces	Area
1	Entrance and moving space	2m Wide Minimum
2	Counters Height width	1.6m-1.8m 0.4m-0.8m

Sr. No	Spaces	Area
3	Working space behind the counters	3m wide
4	Post master Room	9-12m^2
5	Sorting space	2m – 3m per postman
6	Safe and store room	3.0m x 3.0m
7	Waiting and writing space	1.5m x 1.5m
8	Toilets For Male Female	1.5m x 3.0m

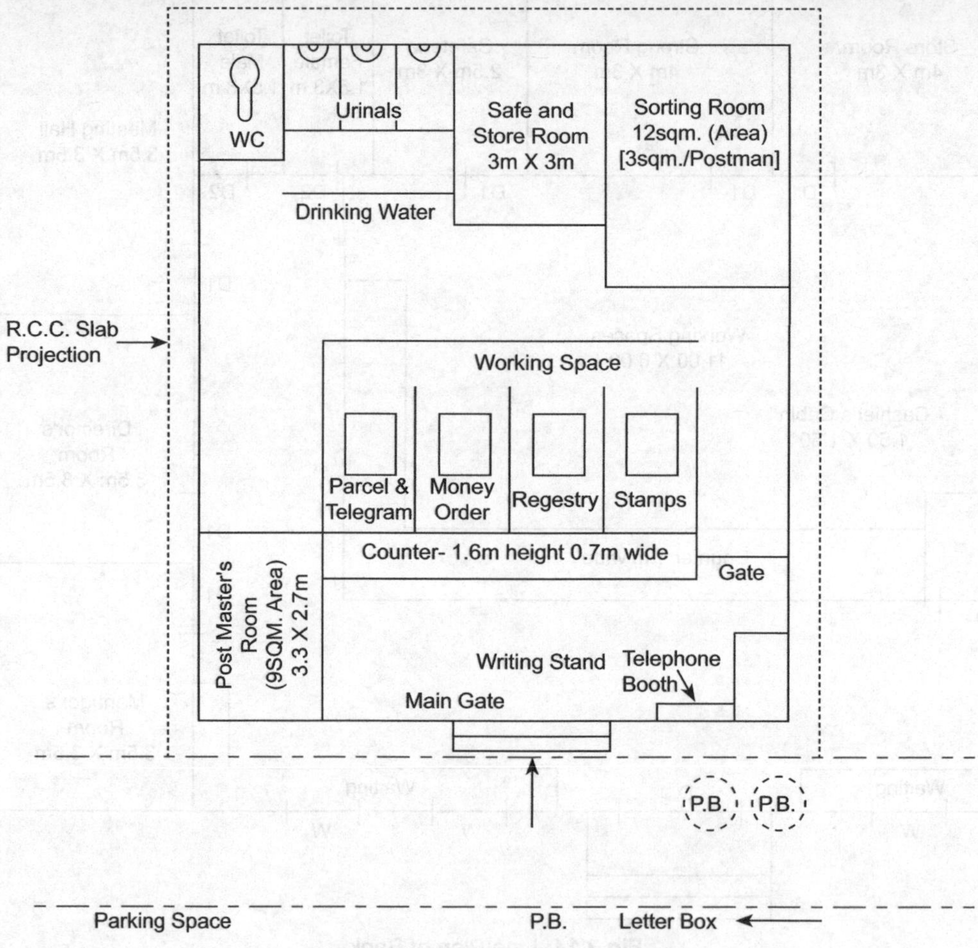

Fig 2.15: Line Plan of Post Office

2.6.6 Requirements of Hostel

Sr. No	Spaces	Area
1	Entrance	3m wide
2	Rooms Single seated	3m x 3m

Sr. No	Spaces	Area
3	Two seated	3.m x 4m
4	Three seated	4.2m x 5m
5	Circulation	1m-2m wide
6	Recreation Hall	2-3m²/head
7	Dining Hall	3-4m²/head
8	Kitchen	2.75m x 3m
9	Pantry	2m x 3m
10	Store	2.75m x 3m
11	Parking space and cycle stand	
12	Sanitary Block	

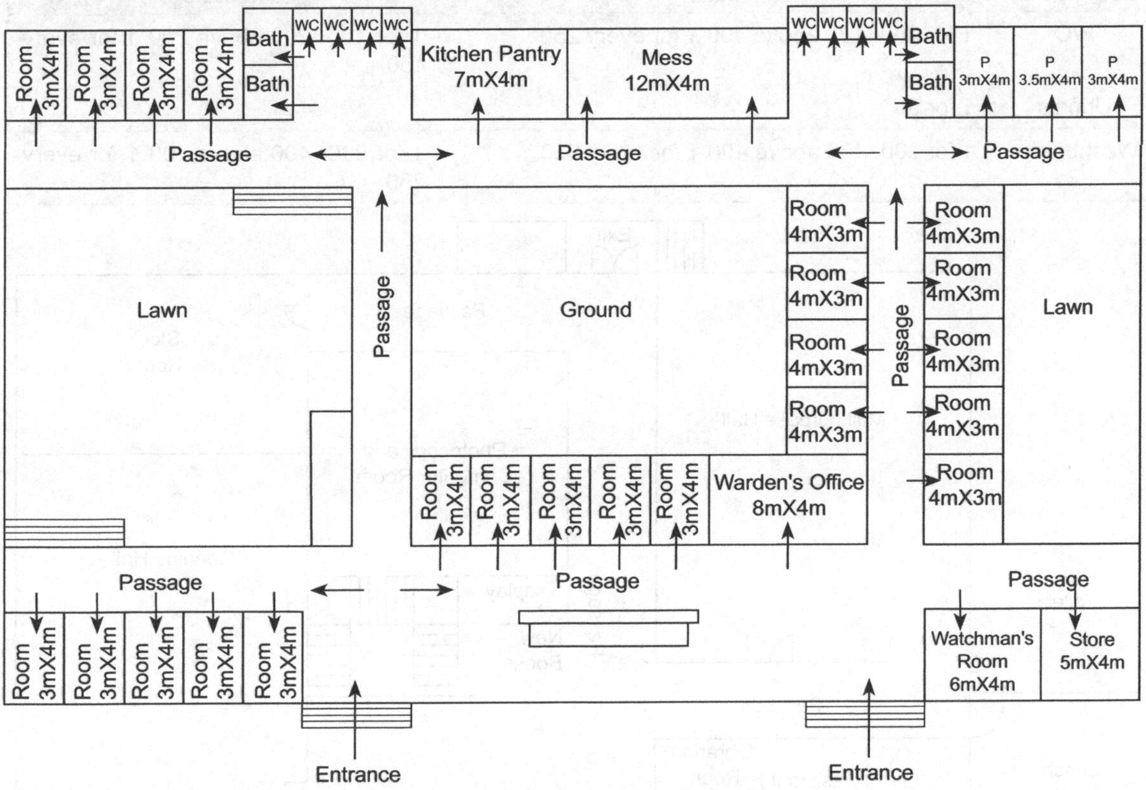

Fig 2.16: Line Plan of Hostel

2.6.7 Requirements of Function Hall and Library

Sr. No	Spaces	Area
1	Entrance and space around Delivery counter	1.8m wide (Minimum)

Sr. No	Spaces	Area
2	Delivery counter	1.6-1.8m height
3	Librarian's Room	9m^2
4	Sections in Library	
	Reference Books Book Section Magazines and News paper Stack Space	20-60 m^2 20-60 m^2 3 m wide (Minimum) 20-40 m^2
5	Circulation	
6	Parking Space and cycle stand	
7	Sanitary block	
Description	**Male**	**Female**
WC	1 for 200 -400 above 400 1 for every 250	1 for 100 -200 above 200 1 for every 150
Urinals	1 for 50	
Washbasins	1 for 200 -400 above 400 1 for every 250	1 for 200 -400 above 400 1 for every 250

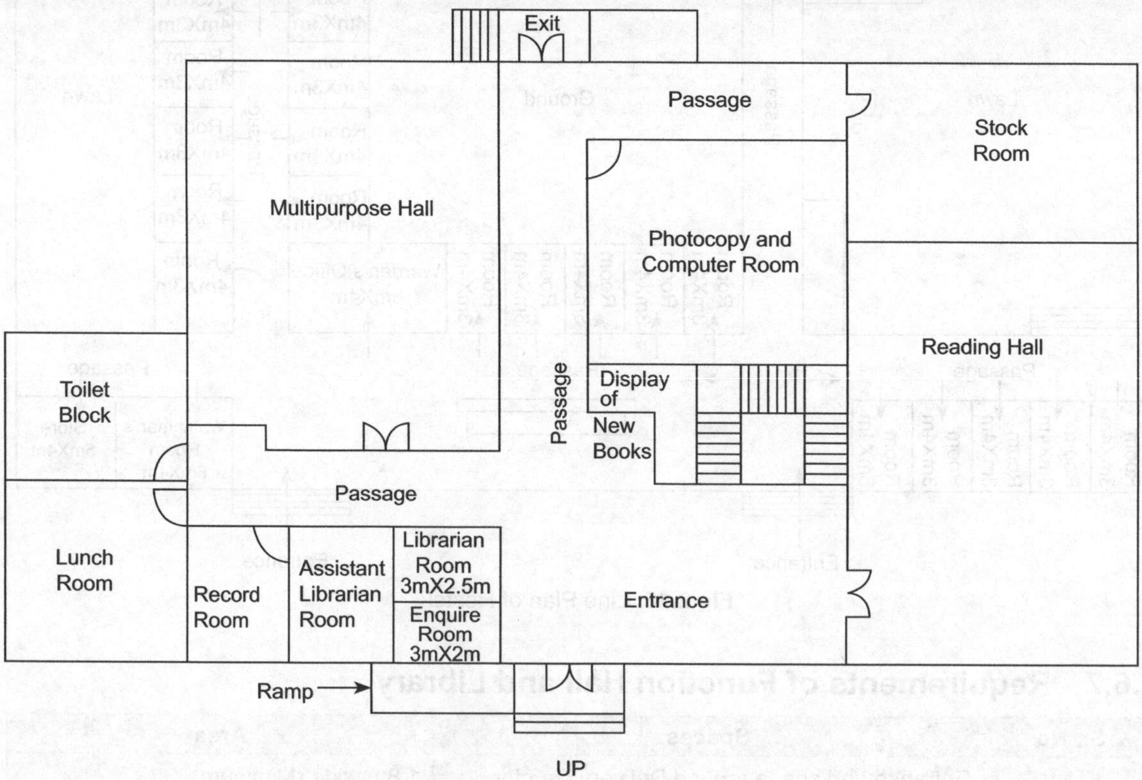

Fig 2.17: Line Plan of Function Hall and Library

EXERCISE

Multiple Choice Questions

1. Principles of Planning is
   (a) Orientation (b) FAR
   (c) Plinth Area (d) Plot Area
2. Following is not a Planning of Principle
   (a) Orientation (b) Grouping
   (c) Privacy (d) Economy
3. Floor Area Ratio is also
   (a) Floor Space Index (b) Floor Space Number
   (c) Floor Space Area (d) Floor Space volume
4. Space area where Carpet can be laid
   (a) Plot Area
   (b) Carpet Area
   (c) Built up Area
   (d) Plinth Area
5. Minimum Dimension of Riser in Residential Building.
   (a) 150mm
   (b) 190mm
   (c) 200mm
   (d) 220mm

Answer to Multiple-Choice Questions				
1. (a)	2. (d)	3. (a)	4. (b)	5. (a)

Answer the Following Questions:

1. Draw Line Plan to suitable Scale IBHK With Staircase and WC and Bathroom.
2. Draw Line Plan to suitable Scale of any building like post office, Hostel, School Building.
3. Measure the room in which you stay and draw line plan of the same.
4. What are Building Bye laws? how they are useful in making the drawing for any new project?
5. What are the principles of planning? Elaborate any five
   (a) Aspect
   (b) Prospect
   (c) Orientation
   (d) Privacy
   (e) Circulation
   (f) Sanitation

REFERENCES

1. Shah. M.G. Kale, CM, Patki, S.Y., Building Drawing, Mcgraw Hill Publishing company Ltd. New Delhi.
2. Sane, Y.S., Planning and Designing of Building, Allied Publishers, New Delhi.
3. Hiraskar G.K, Fundamentals of Town Planning, Dhanpat Rai Publication.
4. Nayak J.B and Prajapati J.A Handbook on Energy Conscious Buildings Publication.
5. Swamy, Kumara; Rao, N, Kameshwara, A., Building Planning and Drawing, Charotar Publication, Anand.

3 Drawing of a Load Bearing Structure

UNIT SPECIFICS

Through this unit we have discussed the following aspects:

To Draw the Plan, Elevation, Section and Details of

- Load Bearing Structure
- Draw and read Plan, Elevation, Section and Other Detailing
- Schedule of Openings and Specifications
- To make Working Drawings
- To draw Foundation Plan

UNIT OUTCOMES

List of outcomes of this unit is as follows:

U3-O1: Understand Load Bearing Structure.

U3-O2: To learn how to Draw Plan Elevation Section and Other Detailing.

U3-O3: To know the Schedule of Openings and Specifications.

U3-O4: Learn to make Working Drawings.

U3-O5: Learn to draw Foundation Plan.

Unit-3 Outcomes	EXPECTED MAPPING WITH COURSE OUTCOMES (1- Weak Correlation; 2- Medium Correlation; 3- Strong Correlation)				
	CO-1	CO-2	CO-3	CO-4	CO-5
U3-O1	2	3	1	1	-
U3-O2	1	3	-	-	-
U3-O3	2	2	2	1	-
U3-O4	3	1	3	3	-
U3-O5	2	2	3	1	-

Scan Me

3.1 LOAD BEARING RESIDENTIAL BUILDING (2 BHK) WITH STAIRCASE

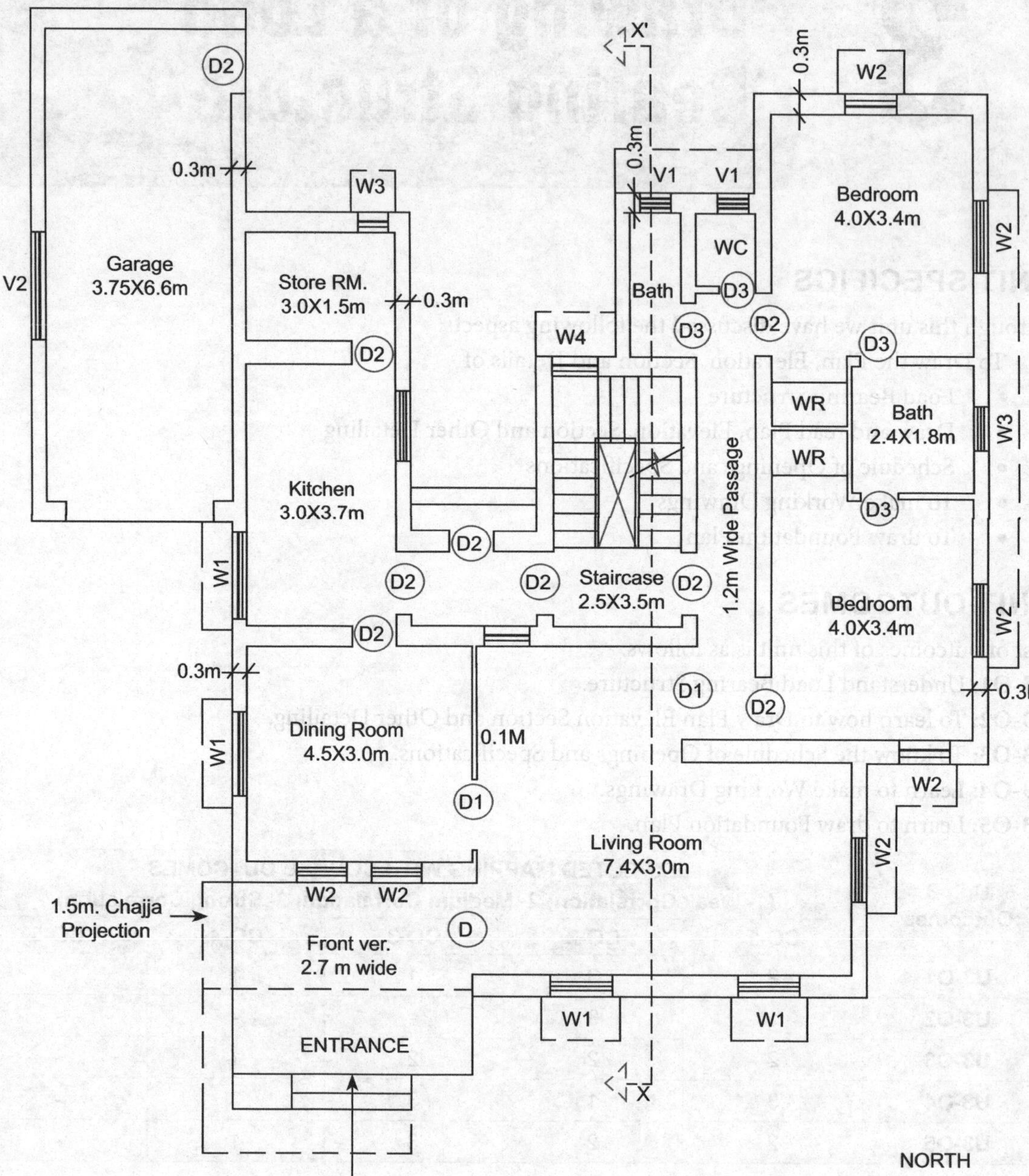

Fig. 3.1: Plan

3.2 DATA DRAWING

3.2.1 Plan

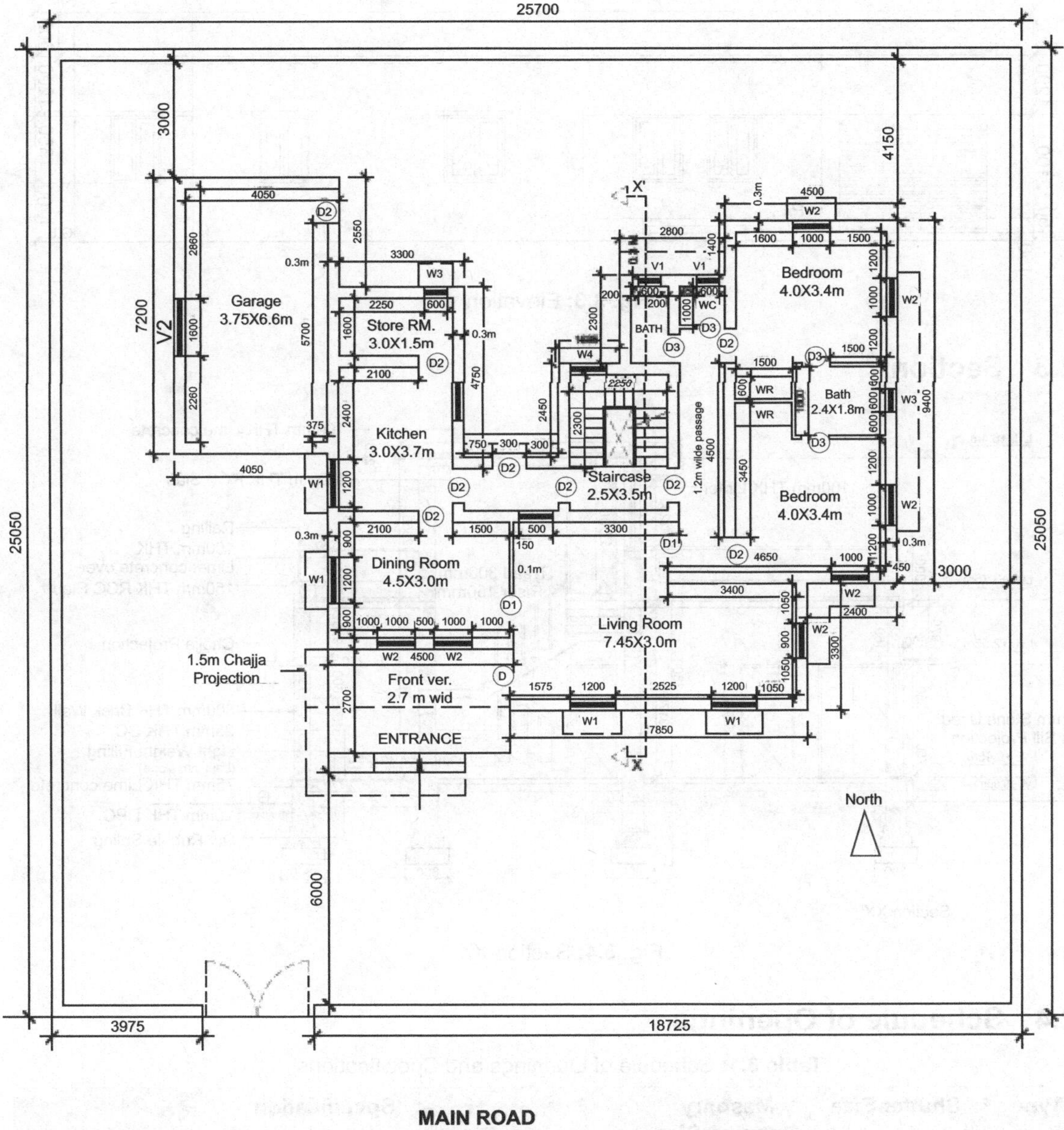

Fig. 3.2: Ground Floor Plan

3.2.2 Elevation

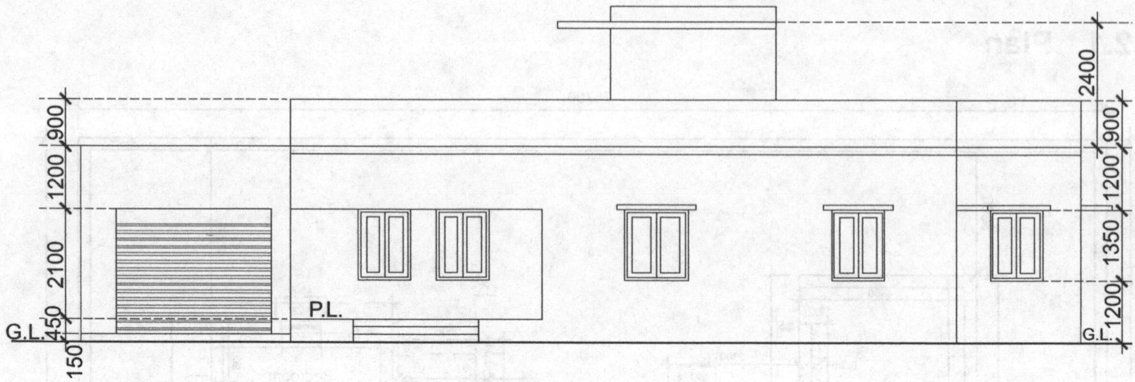

Fig. 3.3: Elevation

3.2.3 Section

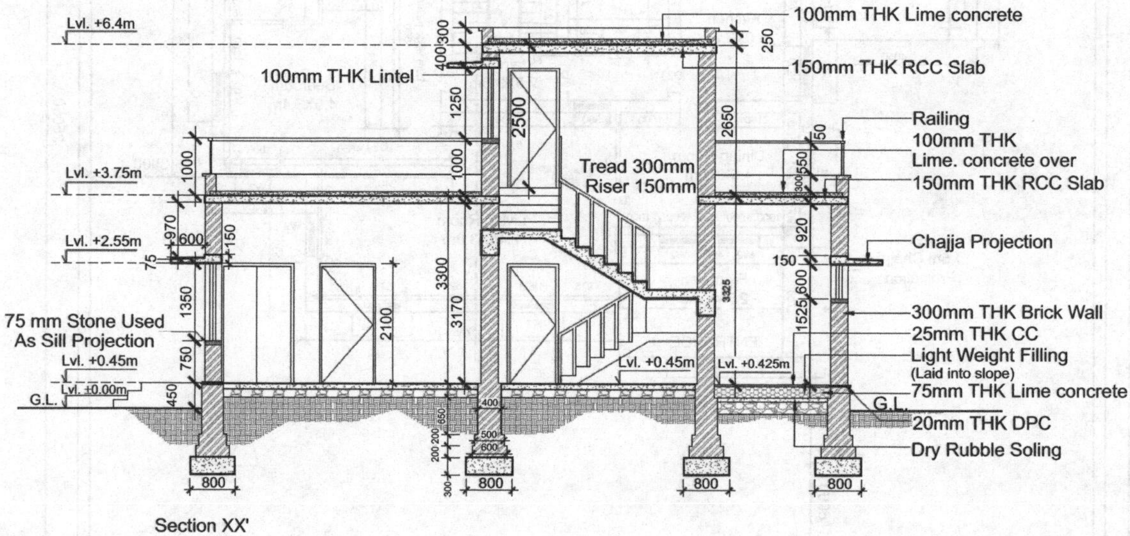

Fig. 3.4: Section XX

3.2.4 Schedule of Openings

Table 3.1: Schedule of Openings and Specifications

Type	Shutter Size	Masonry Opening Size	Specification
D	1100 X 2045	1200 x 2100	Single Shutters Panelled Door shall be teak wood 35 mm thick consisting of T.W Top rail size 100 mm x 35 mm and Bottom rail and Lock Rail 200mm x 35mm insert panel 12mm thick commercial ply faced Teak wood particle board.

Type	Shutter Size	Masonry Opening Size	Specification
			All fittings such as hinges, tower bolts, handles etc. shall be approved make. Iron screws shall be used with iron fittings and brass fittings shall be used with brass fittings.
D1	1000 X 2045	1100 x 2100	Shutters shall be teak wood 35 mm thick Panelled Door. Fittings should be approved before.
D2	900 X 2045	1000 x 2100	Shutters shall be teak wood 35 mm thick Panelled Door. Fittings should be approved before.
D3	750 X 2045	850 x 2100	Shutters shall be teak wood 35 mm thick Panelled Door or Solid PVC Shutter. Fittings should be approved before.
W1	1500 X 1045	1600 x 1100	Wooden windows shutter openable. Fittings should be approved before.
W2	1200 X 1045	1300 x 1100	Wooden windows shutter openable. Fittings should be approved before.
W3	800 X 600	900 x 800	Wooden windows shutter openable. Fittings should be approved before.
W4	900 X 1300	1000 X 1400	Wooden windows shutter openable. Fittings should be approved before.
V1	600 X 600	700 x 800	The louvers of the windows shall consist of 6.35mm thick frosted sheet The louver shall be filled every 75mm in the opening of the windows.
V2	1200 X 600	1300 x 800	The louvers of the windows shall consist of 6.35mm thick frosted sheet glass The louver shall be filled every 75mm in the opening of the windows
ROLLING SHUTTER	3000 X 3000	3000 X 3000	Steel Rolling shutter 18 guage of interlocked laths with shaft, spring encasing box, push and pull operations, inside and outside locking arrangements and all other fittings.

3.2.5 Area Statement

Table 3.2: Statement of Areas

Sr. No.	Room	Length	Breadth	No	Total Area in Sq. M
1	Dining Room	4.5	3	1	13.5
2	Drawing Room	7.45	3.5	1	26.075
3	Bedroom	4	3.4	1	13.6
4	Bedroom	4	3.4	1	13.6
5	Storeroom	3	1.5	1	4.5
6	Kitchen	3	3.7	1	11.1
7	Toilet	2.4	1.8	1	4.32

Sr. No.	Room	Length	Breadth	No	Total Area in Sq. M
8	Bath	1.5	2	1	3
9	W.C	1.2	1	1	1.2
10	Staircase	2.55	2.45	1	6.2475
11	Garage	3.75	6.6	1	24.75
				Total	121.89

3.2.6 Construction Notes with Specification

- **Foundation and Plinth:** Foundation and Plinth of the load Bearing structure shall be 1st class brick work of 1:6 cement mortar over lime concrete or Lime Mortar or 1:4:8 cement concrete.
- **Damp Proof Course (DPC)** :D.P.C shall be 20mm thick cement concrete 1:1.5:3, mixed with 1kg of Impermo with 1kg bag of cement or other standard water proofing materials as specified and painted with two coats of bitumen.
- **Superstructure:** Superstructure shall be of first-class brick work with 1:6 Lime mortar or cement concrete. Lintels over doors and windows shall be of R.C.C.
- **Roofing:** Roofing shall be Reinforced Cement Concrete Slab and lime concrete terracing above, supported over R.S Joists or R.C.C Beam as required. Height of the room will be 3.5-3.7m
- **Flooring:** Drawing room and Dining room floor shall be mosaic (terrazzo), bathroom and W.C flooring and Dado shall be mosaic(terrazzo). Flooring of bedrooms and other rooms shall be coloured and polished 2.5cm layer of cement concrete over 7.5cm layer of lime concrete.
- **Finishing:** 12 mm cement lime plaster in 1:1:6 ratio on Inside and outside brick wall, Drawing room, Dining room and Bedrooms – inside wall shall be painted with distemper and others in side white washed 3coats. Outside wall shall be coated with two coats of coloured distemper over two coats of Birla white putty.
- **Doors and Windows:** Chaukhats or Frames shall be seasoned teak wood, shutters shall be teak wood 40 mm thick panelled glazed or partly panelled and partly glazed as required. All fittings such as hinges, tower bolts, handles etc. shall be approved make. Iron screws shall be used with iron fittings and brass fittings shall be used with brass fittings. Doors and windows shall be varnished or polished with two coats with high class enamel paint over one coat of priming. Window shall be provided with iron gratings or grills.
- **Miscellaneous:** Cast iron or of PVC Rainwater pipes shall be provided. Sanitary, Water fittings and Electricals Installations shall be 1st class quality or approve make. 1 meter wide 7.5cm thick cement concrete in 1:3:6 apron/plinth protection shall be provided around the building.

3.2.7 Riser and Tread in Residential Building

Riser : 175-185 mm

Tread : 250-300 mm

Riser and Tread in Public Building:

Riser : 150-170 mm

Tread : 270-300 mm

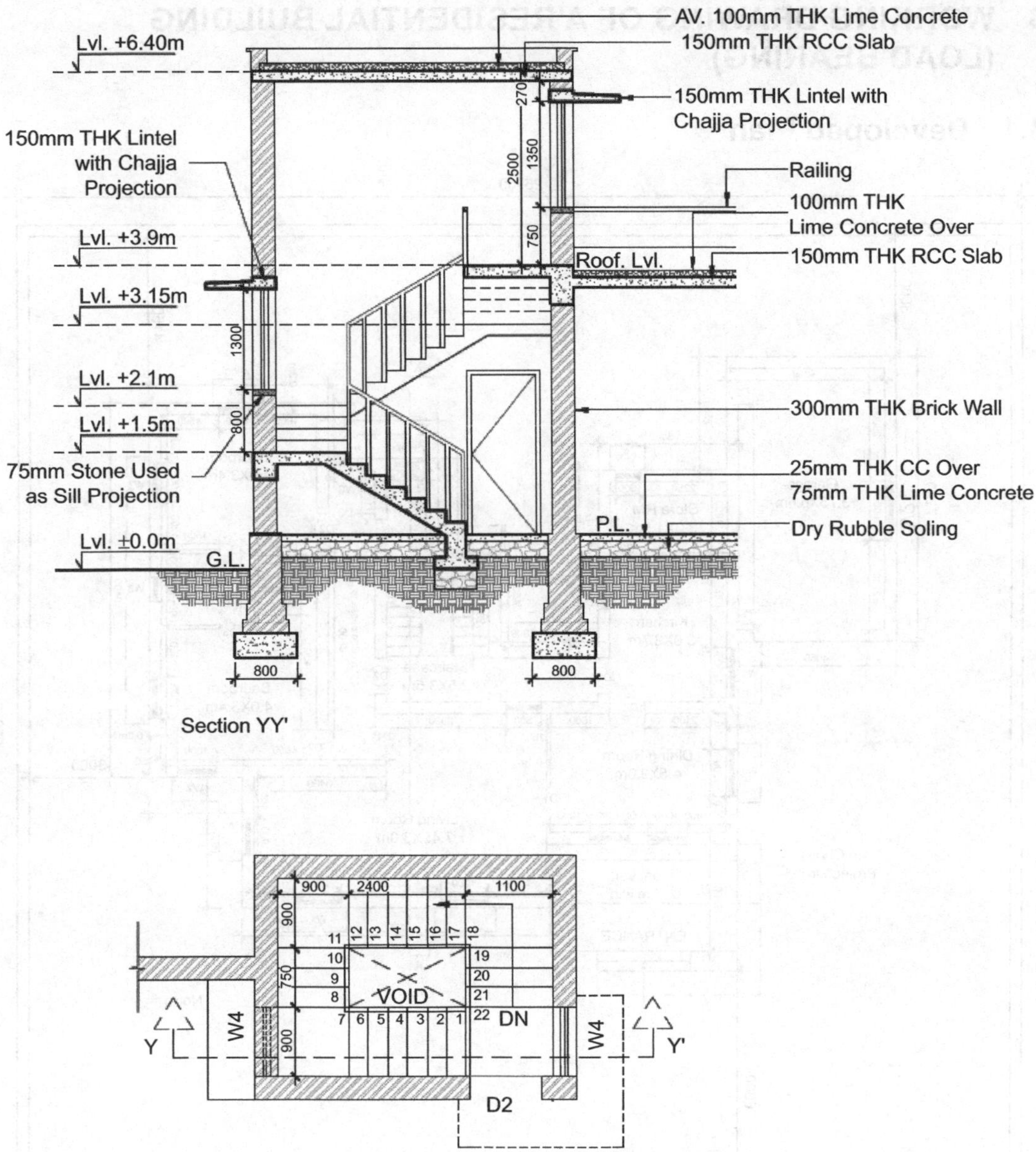

Fig. 3.5: Plan and Section of a Residential Staircase

3.3 WORKING DRAWING OF A RESIDENTIAL BUILDING (LOAD BEARING)

3.3.1 Developed Plan

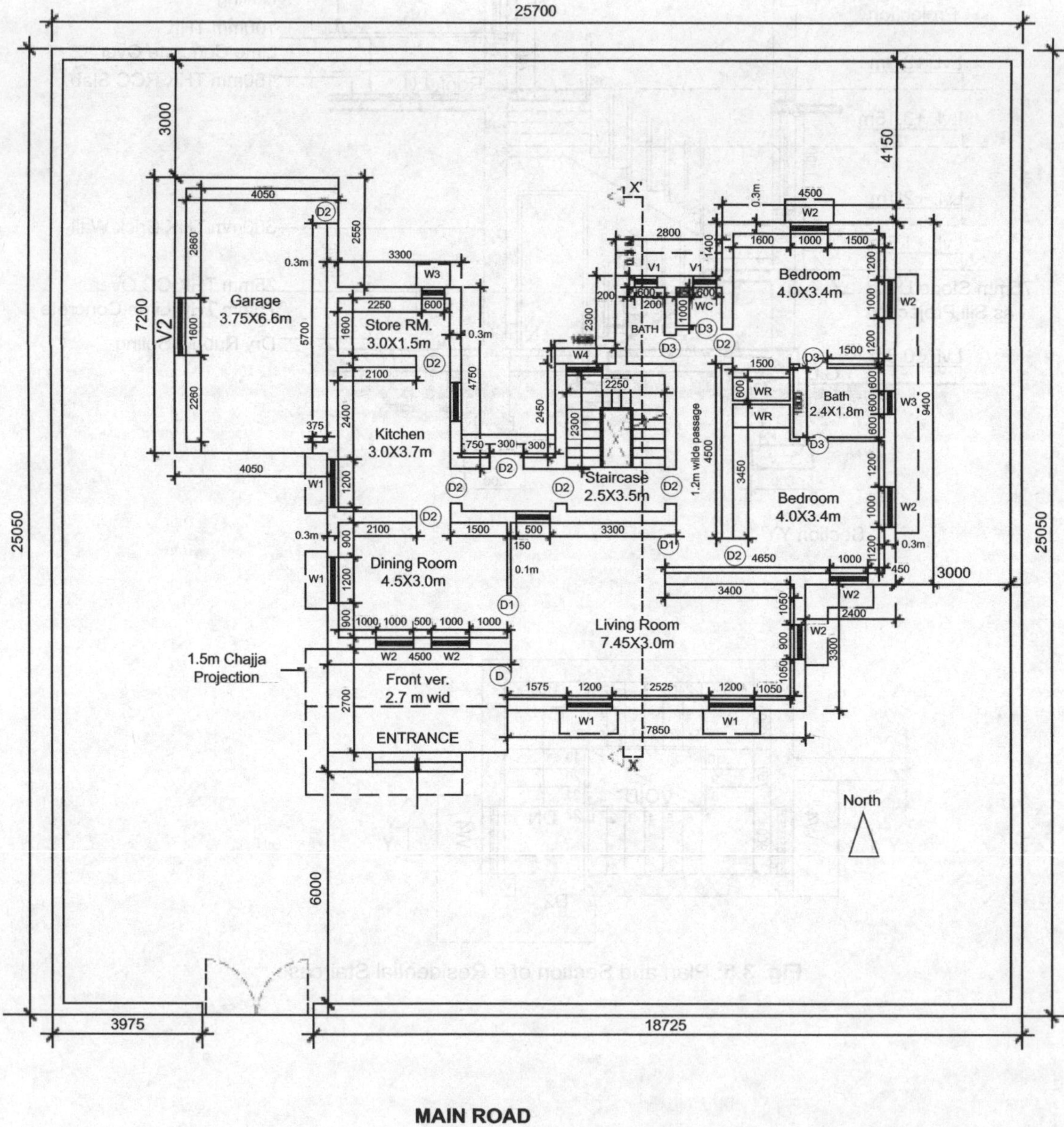

Fig. 3.6: Developed Plan

3.2.3 Elevation

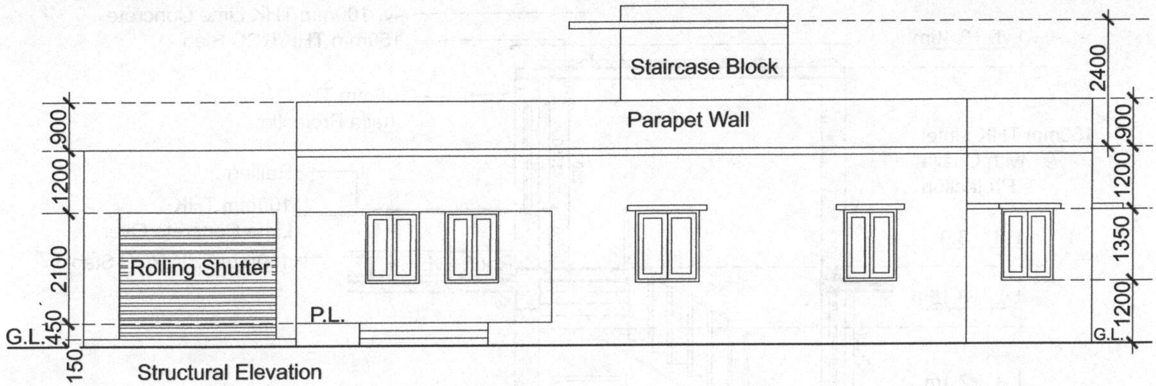

Fig. 3.7: Elevation

3.2.4 Section

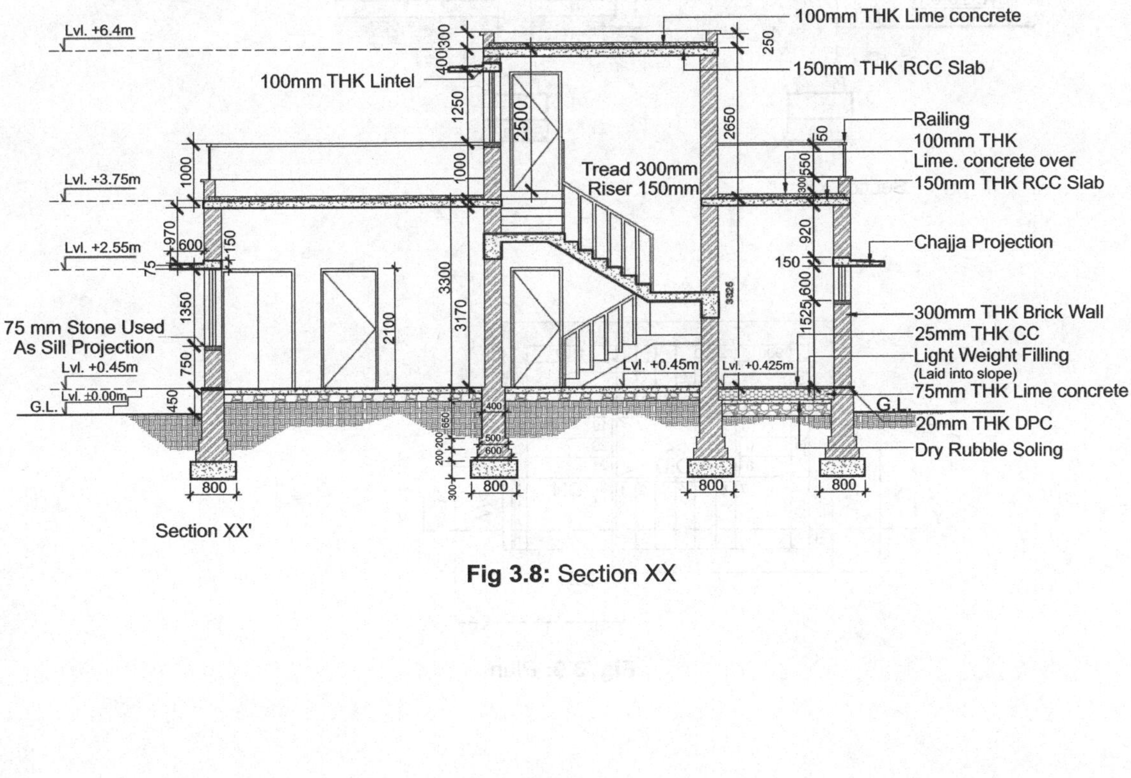

Fig 3.8: Section XX

3.2.5 Working Drawing of Staircase

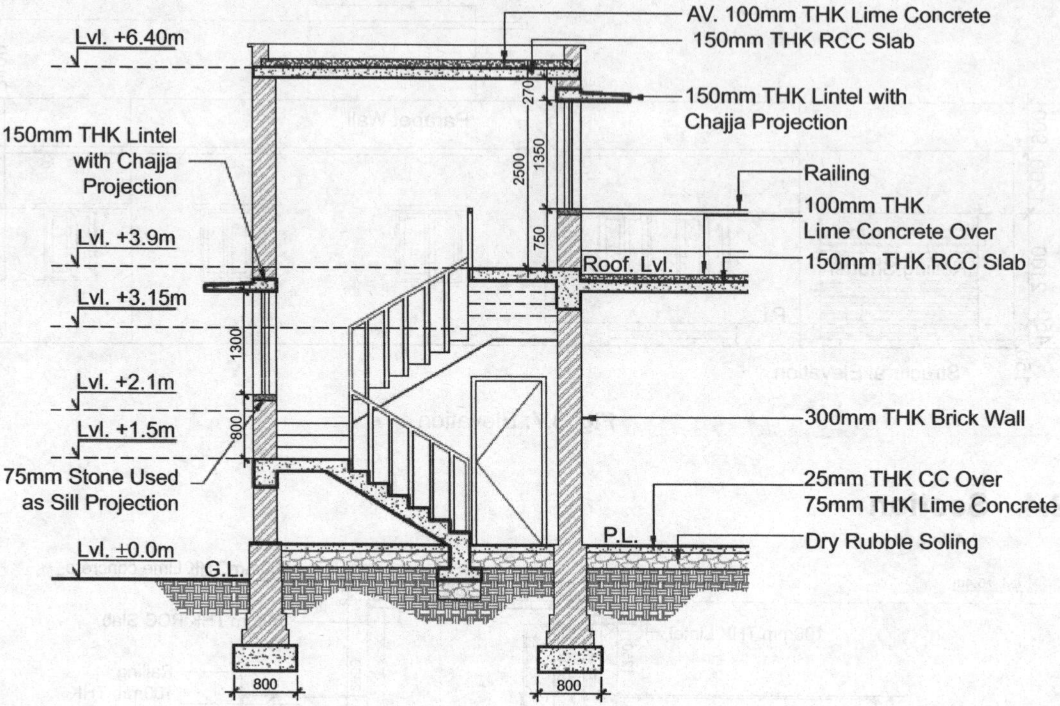

Section YY'

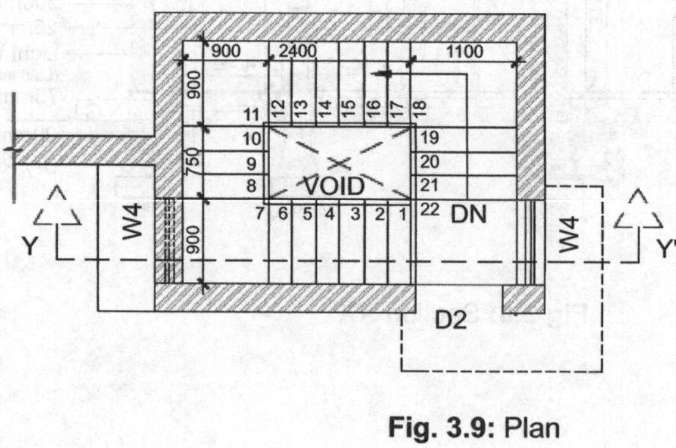

Fig. 3.9: Plan

3.4 FOUNDATION PLAN OF THE ABOVE LOAD BEARING STRUCTURE

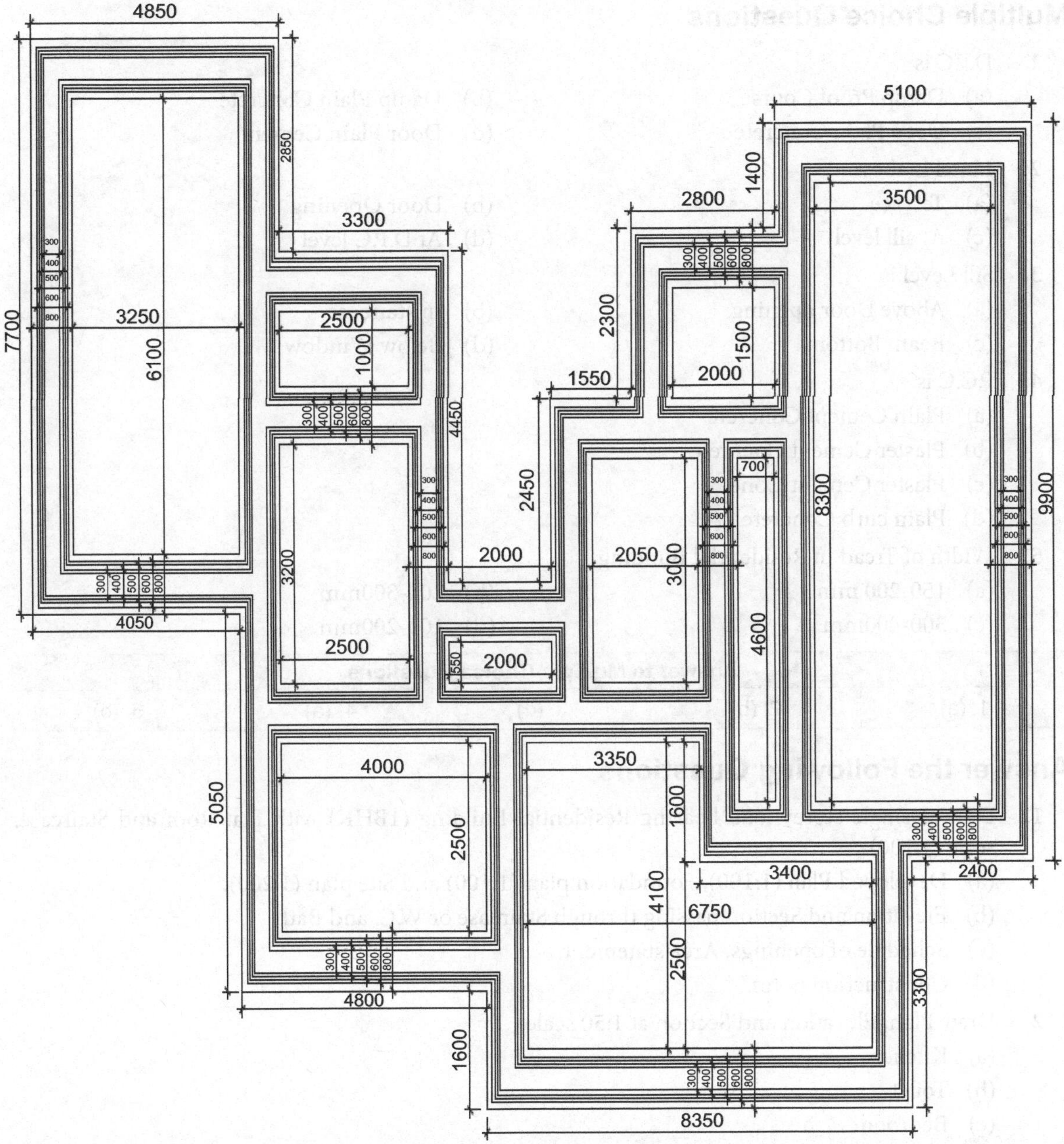

Fig 3.10: Foundation Plan

EXERCISE

Multiple Choice Questions

1. D.P.C is
   (a) Damp Proof Course
   (b) Damp Plain Concrete
   (c) Deep Plain Concrete
   (d) Door Plain Cement
2. Lintel is above
   (a) Terrace
   (b) Door Opening
   (c) At sill level
   (d) At D.P.C level
3. Sill Level is
   (a) Above Door opening
   (b) In staircase
   (c) Beam Bottom
   (d) Below Window
4. P.C.C is
   (a) Plain Cement Concrete
   (b) Plaster Cement Concrete
   (c) Plaster Cement Cone
   (d) Plain curb Concrete
5. Width of Tread in Residential Building
   (a) 150-200 mm
   (b) 200-300mm
   (c) 300-400mm
   (d) 100-200mm

Answer to Multiple-Choice Questions				
1. (a)	2. (b)	3. (d)	4. (a)	5. (b)

Answer the Following Questions

1. Draw a single storey load bearing Residential Building (1BHK) with Flat Roof and Staircase. scale (1:100)
   (a) Developed Plan (1:100), Foundation plan (1:100) and Site plan (1:200).
   (b) Elevation and Section passing through Staircase or W.C. and Bath
   (c) Schedule of openings, Area statement
   (d) Construction notes.
2. Draw Plan, Elevation and Section at 1:50 scale
   (a) Kitchen
   (b) Toilet
   (c) Bedroom
3. Draw Section of the footing for 300mm thick brick wall in 1:10 scale?
4. Draw Freehand
   (a) Section of the wall 300 mm thick with door opening
   (b) Section of the wall with window opening.

5. Draw Staircase
Tread 300mm
Riser 150mm

REFERENCES

1. Dutta B.N, Estimating and Costing, Dutta and Company Publishing 1986
1. S.V Deodhar, Building Drawing, New Vrinda Publishing House ,1992
2. Sane, Y.S., Planning and design of Building, Allied Publishers, New Delhi,1964.
3. Swamy, Kumara; Rao, N, Kameshwara, A, Building Planning and Drawing, Charotar Publication, Anand.
4. Shah. M.G. Kale, CM, Patki, S.Y., Building Drawing, Mcgraw Hill Publishing company Ltd. New Delhi.
5. Public Works Department, Govt. of Maharashtra, State Schedule of Rate 2022-23
6. Central Public Works Department, Manual on Door and Window Detail for residential Building, CPWD 2006

4 Drawing of Framed Structure

UNIT SPECIFICS

Through this unit the following aspects are discussed:

To Draw the Plan, Elevation, Section and Details of

- Frame Structure
- Draw and read Plan, Elevation, Section and Other Detailing
- Schedule of Openings and Specifications
- To make Working Drawings
- To draw Foundation Plan

UNIT OUTCOMES

List of outcomes of this unit is as follows:

U4-O1: Describe Frame Structure

U4-O2: Draw Plan Elevation Section and Other Detailing

U4-O3: Schedule of Openings and Specifications

U4-O4: To make Working Drawings

U4-O5: To draw Foundation Plan using CAD

Unit-4 Outcomes	EXPECTED MAPPING WITH COURSE OUTCOMES (1- Weak Correlation; 2- Medium Correlation; 3- Strong Correlation)				
	CO-1	CO-2	CO-3	CO-4	CO-5
U4-O1	2	2	2	3	-
U4-O2	2	2	1	3	-
U4-O3	1	1	1	2	-
U4-O4	2	1	1	3	-
U4-O5	1	-	-	2	-

4.1 DOUBLE STOREY FRAME STRUCTURE (G+1) RESIDENTIAL BUILDING (2 BHK) WITH STAIRCASE

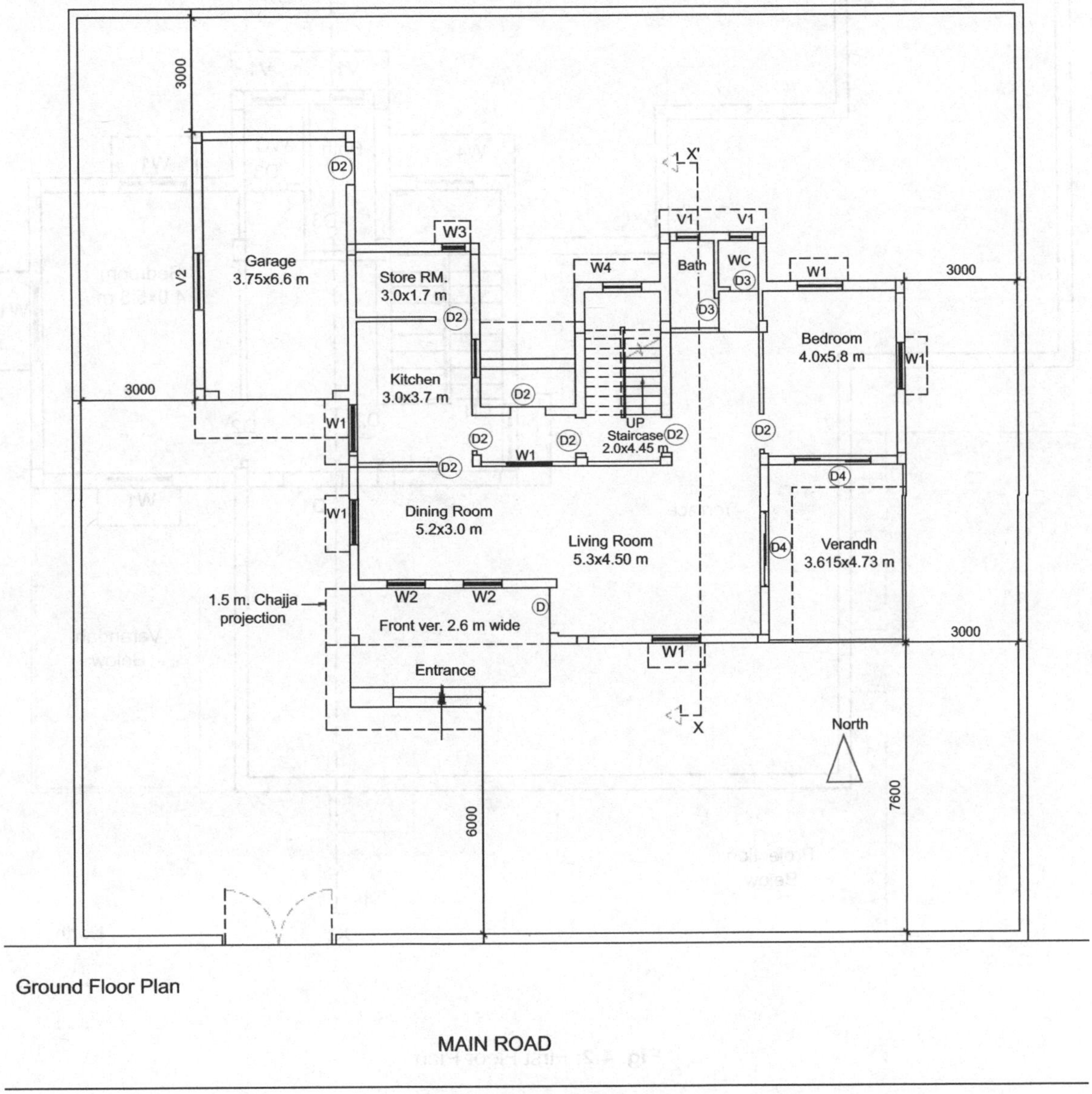

Fig. 4.1: Ground Floor Plan

Scan Me

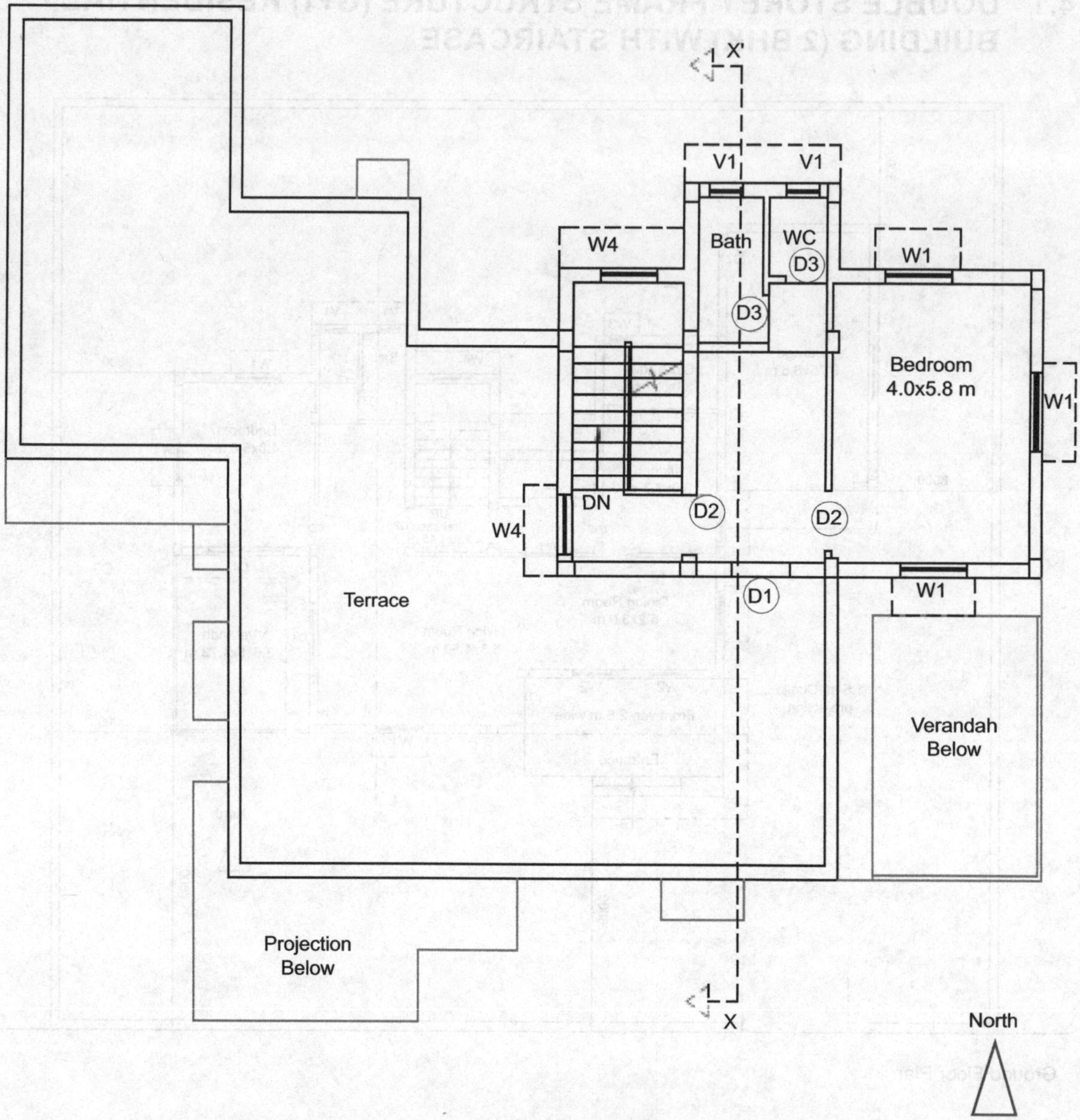

Fig. 4.2: First Floor Plan

4.2 DATA DRAWING –DEVELOPED PLAN, ELEVATION, SECTION, SITE PLAN AND DETAILS:

4.2.1 Plan

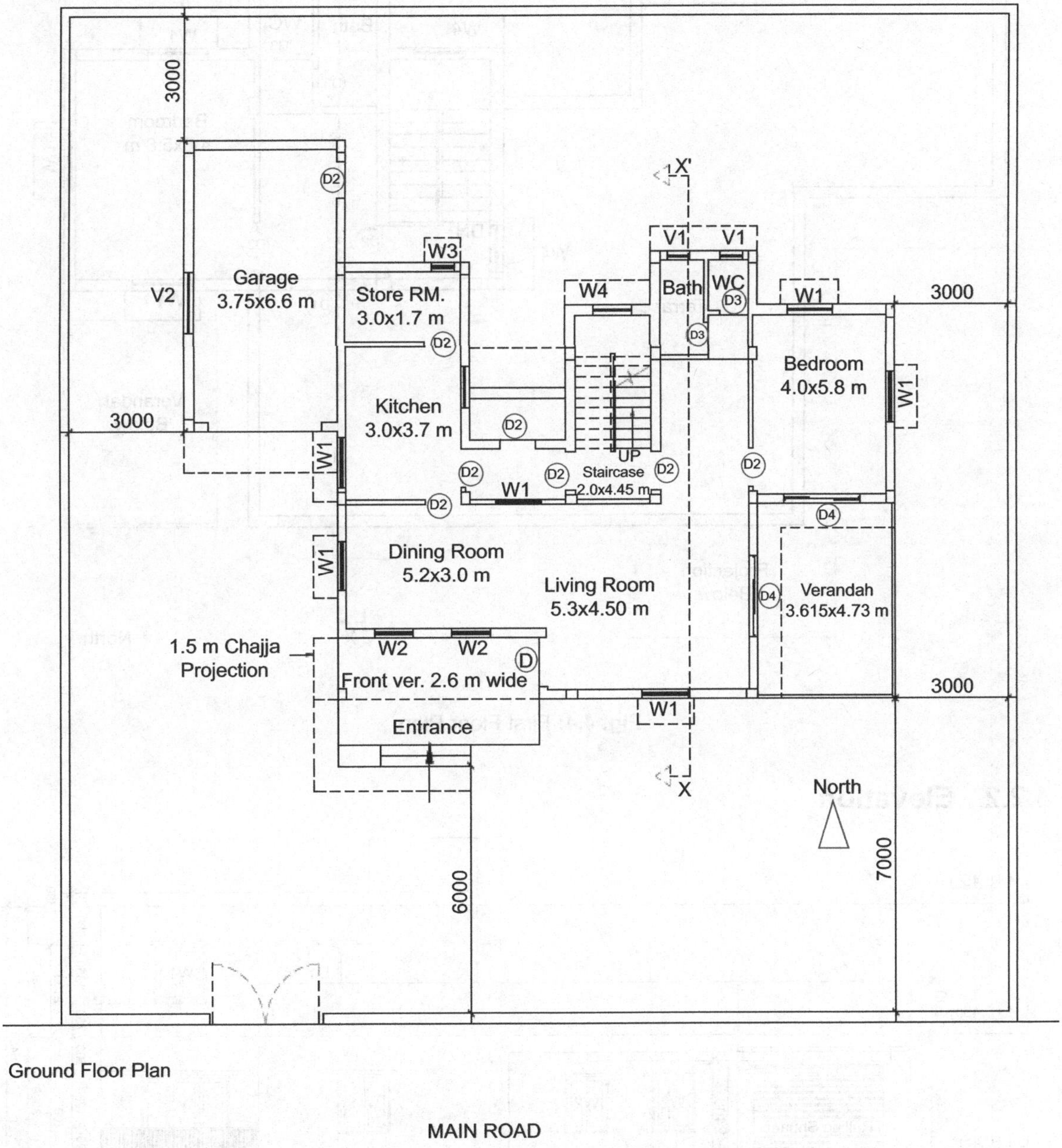

Fig. 4.3: Ground Floor Plan

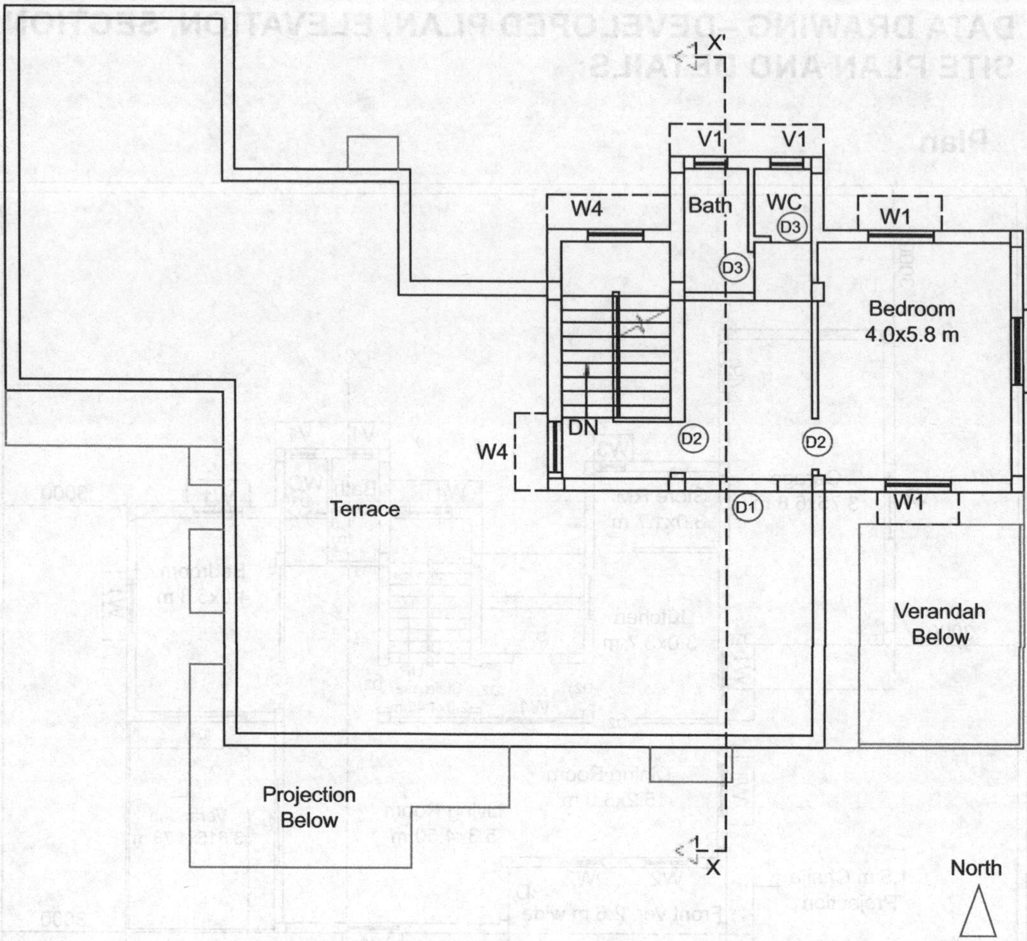

Fig. 4.4: First Floor Plan

4.2.2 Elevation

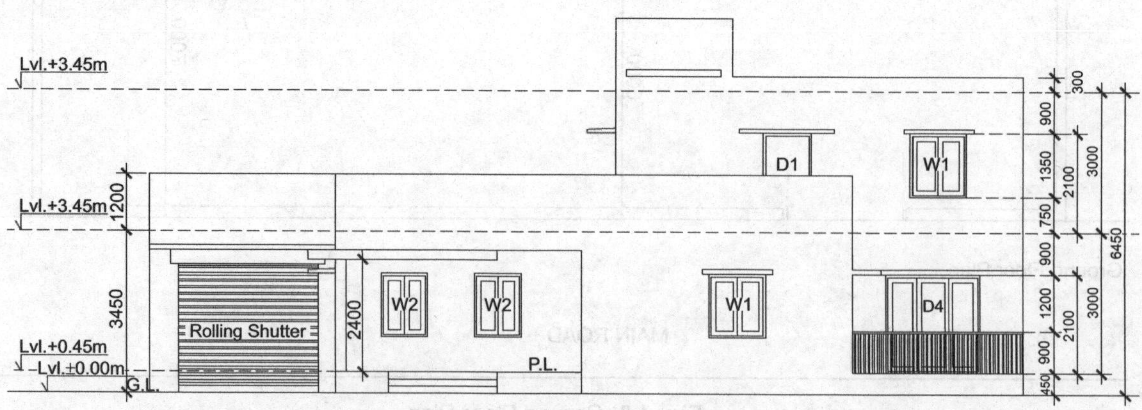

Fig. 4.5: Elevation

4.2.3 Section

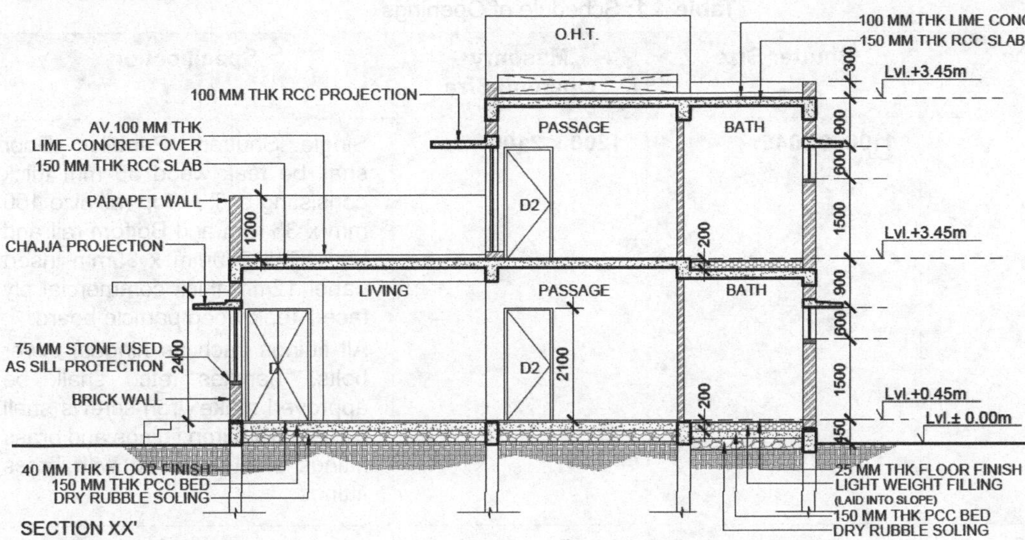

Fig. 4.6: Section XX

4.2.4 Site Plan

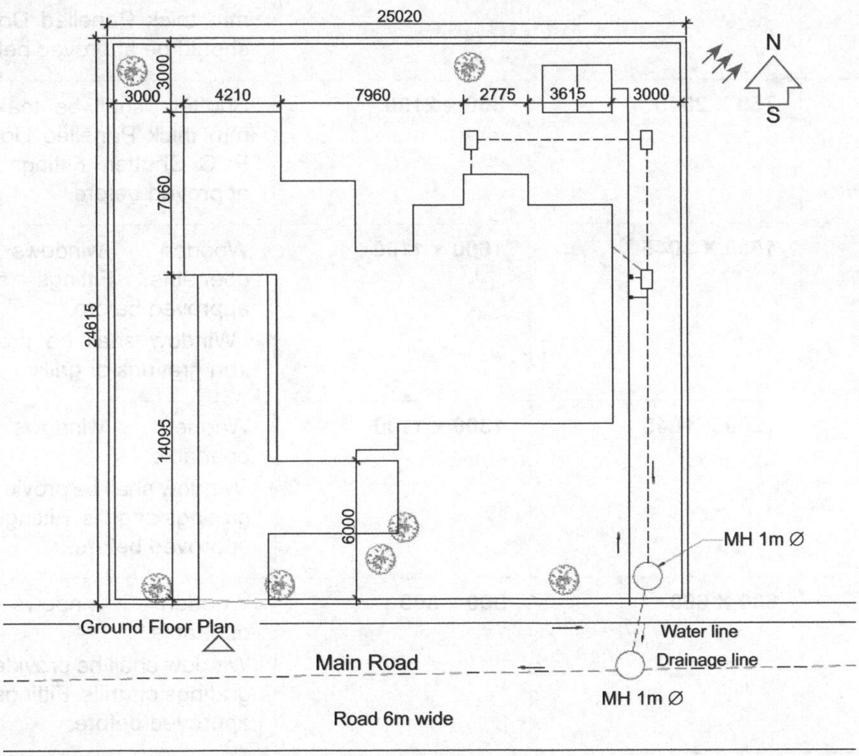

Fig. 4.7: Site Plan

4.2.5 Schedule of Openings

Table 4.1: Schedule of Openings

Type	Shutter Size	Masonry Opening Size	Specification
D	**1100 X 2045**	**1200 x 2100**	Single Shutters Panelled Door shall be teak wood 35 mm thick consisting of T.W Top rail size 100 mm x 35 mm and Bottom rail and Lock Rail 200mm x 35mm insert panel 12mm thick commercial ply faced Teak wood particle board. All fittings such as hinges, tower bolts, handles etc. shall be approved make. Iron screws shall be used with iron fittings and brass fittings shall be used with brass fittings.
D1	**1000 X 2045**	**1100 x 2100**	Shutters shall be teak wood 35 mm thick Panelled Door. Fittings should be approved before.
D2	**900 X 2045**	**1000 x 2100**	Shutters shall be teak wood 35 mm thick Panelled Door. Fittings should be approved before.
D3	**750 X 2045**	**850 x 2100**	Shutters shall be teak wood 35 mm thick Panelled Door or Solid PVC Shutter. Fittings should be approved before.
W1	**1500 X 1045**	**1600 x 1100**	Wooden windows shutter openable. Fittings should be approved before. Window shall be provided with iron gratings or grills.
W2	**1200 X 1045**	**1300 x 1100**	Wooden windows shutter openable. Window shall be provided with iron gratings or grills. Fittings should be approved before.
W3	**800 X 600**	**900 x 800**	Wooden windows shutter openable. Window shall be provided with iron gratings or grills. Fittings should be approved before.

Type	Shutter Size	Masonry Opening Size	Specification
W4	**900 X 1300**	**1000 X 1400**	Wooden windows shutter openable. Window shall be provided with iron gratings or grills. Fittings should be approved before.
V1	**600 X 600**	**700 x 800**	The louvers of the windows shall consist of 6.35mm thick frosted sheet glass or5mm thick acrylic sheet 125 to 150 mm in width. The louver shall be filled every 75mm in the opening of the windows.
V2	**1200 X 600**	**1300 x 800**	The louvers of the windows shall consist of 6.35mm thick frosted sheet glass or5mm thick acrylic sheet 125 to 150 mm in width. The louver shall be filled every 75mm in the opening of the windows
ROLLING SHUTTER	**3000 X 3000**	**3000 X 3000**	Steel Rolling shutter 18 guage of interlocked laths complete with shaft, spring encasing box, push and pull operations, inside and outside locking arrangements and all other fittings.

4.2.6 Area Statement

Table 4.2: Area Statement

Sr. No.	Room	Length	Breadth	No	Total Area in Sq. M
1	Dining Room	5.2	3	1	15.6
2	Drawing Room	5.3	4.5	1	23.85
3	Bedroom	5.8	4	1	23.2
4	Storeroom	3	1.7	1	5.1
5	Kitchen	3	3.7	1	11.1
6	Bath	1.5	2.3	1	3.45
7	W.C	1.2	1	1	1.2
8	Passage	2.5	3.5	1	8.75

Sr. No.	Room	Length	Breadth	No	Total Area in Sq. M
9	Staircase	3.35	2.45	1	8.21
10	Garage	3.75	6.6	1	24.75
				Total	**125.21**

4.2.7 Construction Notes with Specification

- **Foundation and Plinth:** Foundation and Plinth shall be 1st class brick work of 1:6 cement mortar over 1:4:8 cement concrete.
- **Damp Proof Course (DPC)** :D.P.C shall be 20mm thick cement concrete 1:11/2:3, mixed with 1kg of Impermo per bag of cement or other standard water proofing materials as specified and painted with two coats of bitumen.
- **Superstructure:** superstructure shall be of first-class brick work with 1:6 cement mortar. Lintels over doors and windows shall be of R.C.C.
- **Roofing:** Roofing shall be R.C.C Slab lime concrete terracing above, supported over R.S Joists or R.C.C Beam as required. Height of the room will be 3.5-3.7m
- **Flooring:** Drawing room and Dining room floor shall be mosaic (terrazzo), bathroom and W.C floor and Dado shall be mosaic(terrazzo). Floors of bedrooms shall be coloured and polished of 2.5cm cement concrete over 7.5cm lime concrete. Floor of other rooms shall be 2.5cm cement concrete over 7.5 cm lime concrete polished.
- **Finishing:** Inside and outside shall be of 12 mm cement lime plaster 1:1:6, Drawing room, Dining room and Bedrooms – inside wall shall be distempered and others inside white washed 3coats. Outside shall be coloured distemper two coats over two coats of Birla white putty.
- **Doors and Windows:** Chaukhats or Frames shall be seasoned teak wood, shutters shall be teak wood 400mm thick panelled glazed or partly panelled and partly glazed as required. All fittings shall be of brass. Doors and windows shall be varnished or polished with two coats with high class enamel paint over one coat of priming. Window shall be provided with iron gratings or grills.
- **Miscellaneous:** Rainwater pipes of cast iron or of PVC shall be provided and finished painted. Building shall be provided with 1st class Sanitary, Water fittings and Electricals Installations. 1 meter wide 7.5cm thick cement concrete in 1:3:6 apron/plinth protection shall be provided around the building.

4.2.8 Riser and Tread in Residential Building

Riser: 175-185 mm

Tread: 250-300 mm

Riser and Tread in Public Building:

Riser: 150-170 mm

Tread: 270-300 mm

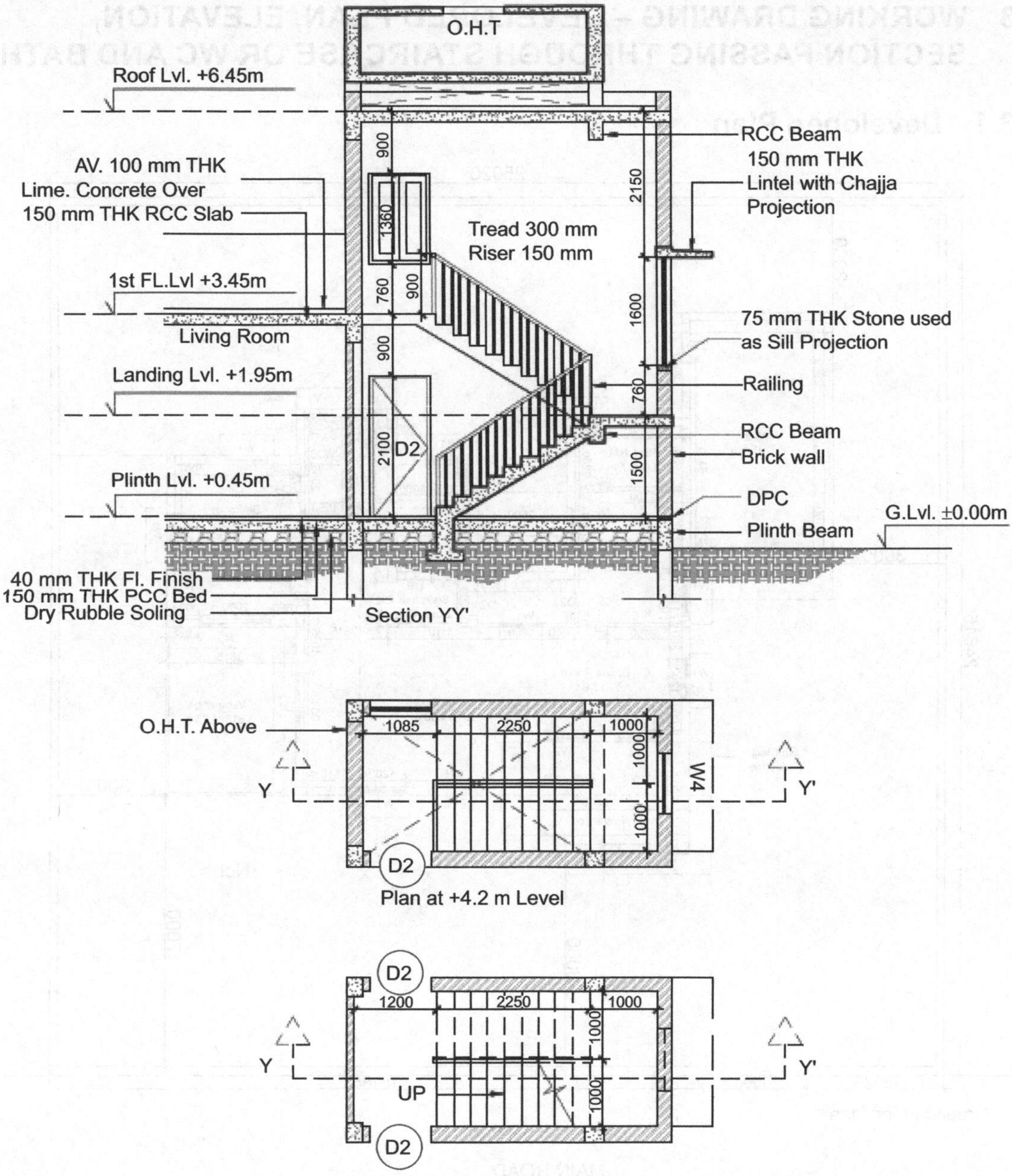

Fig. 4.8: Staircase Plan of Residential Staircase

Scan Me

4.3 WORKING DRAWING – DEVELOPED PLAN, ELEVATION, SECTION PASSING THROUGH STAIRCASE OR WC AND BATH

4.3.1 Developed Plan

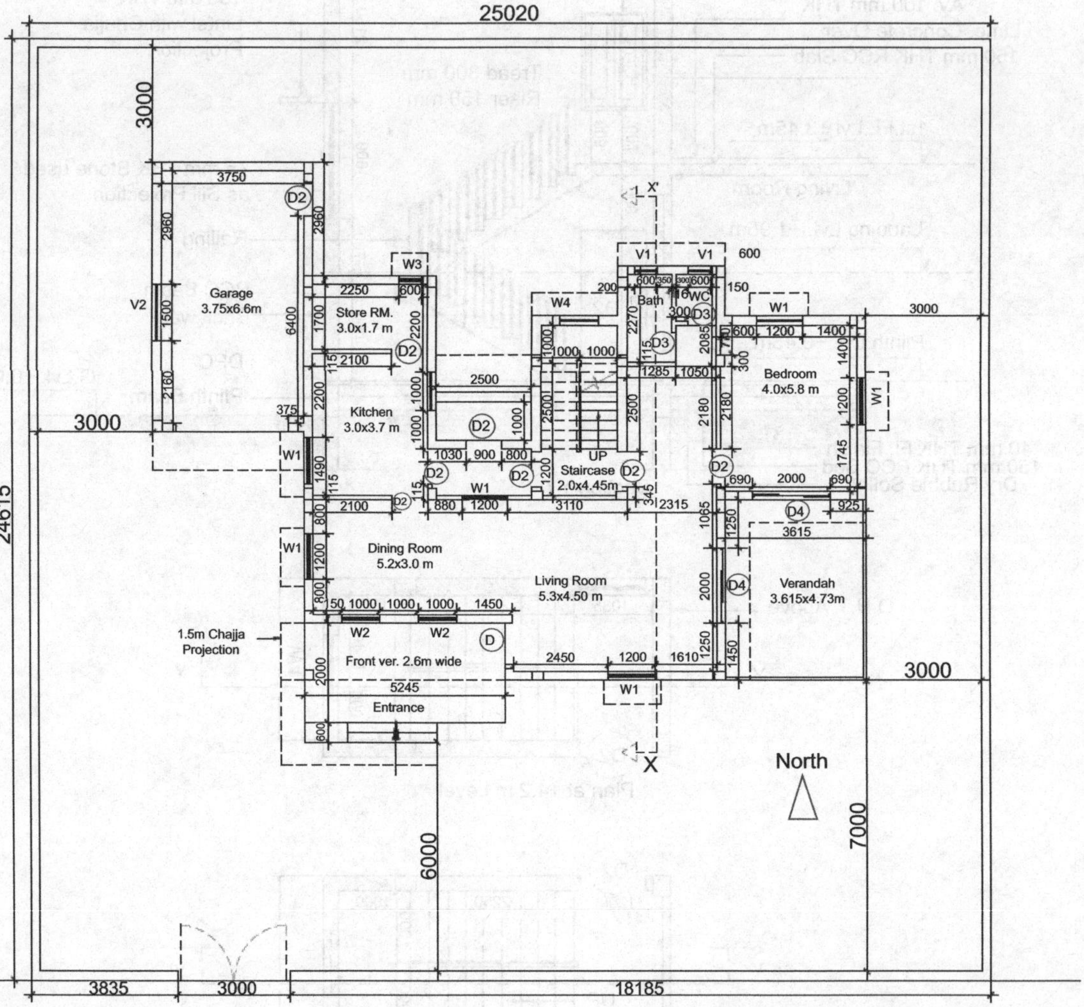

Ground Floor Plan

MAIN ROAD

Fig. 4.9: Ground Floor Plan

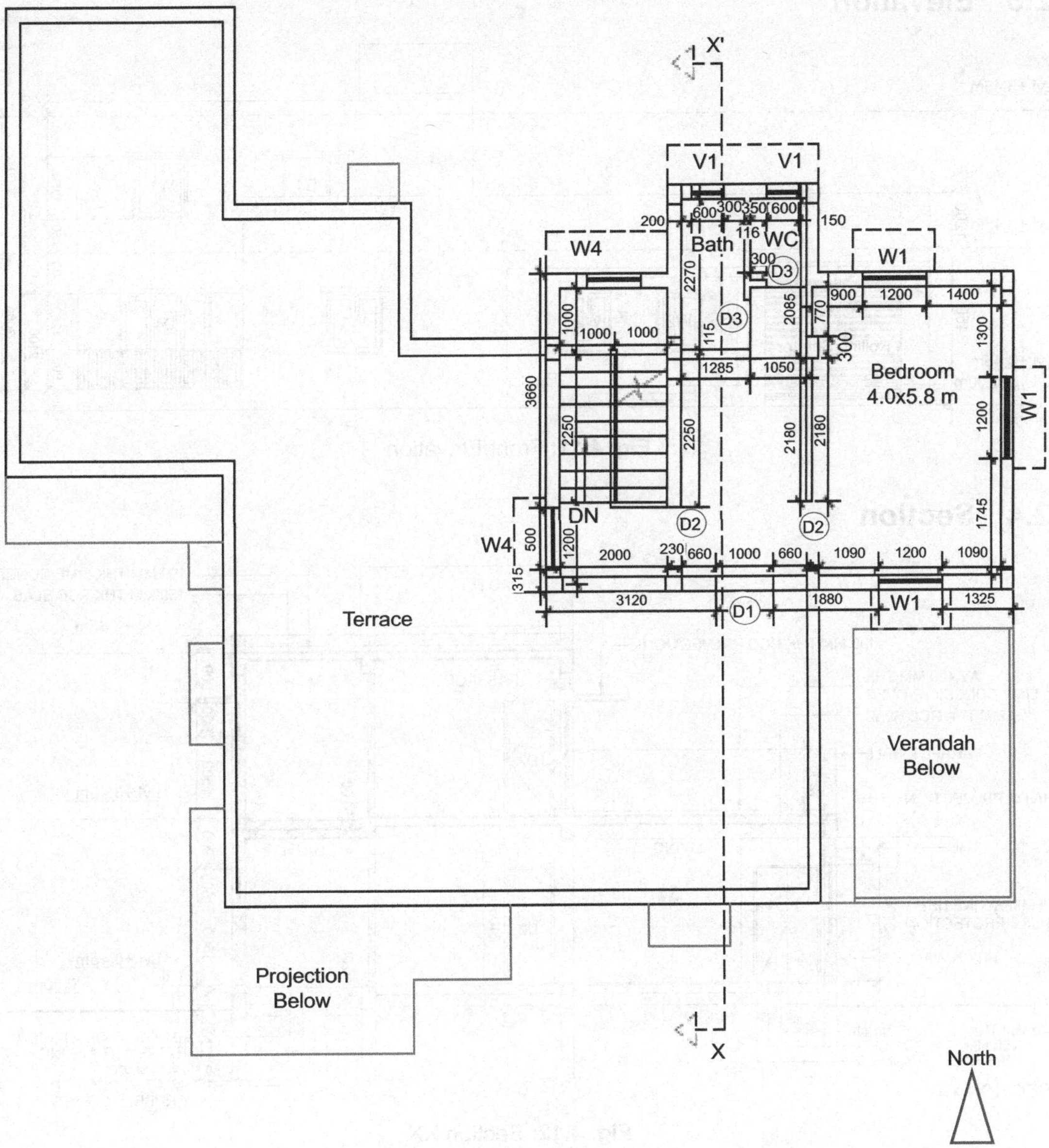

Fig. 4.10: First Floor Plan

4.2.3 Elevation

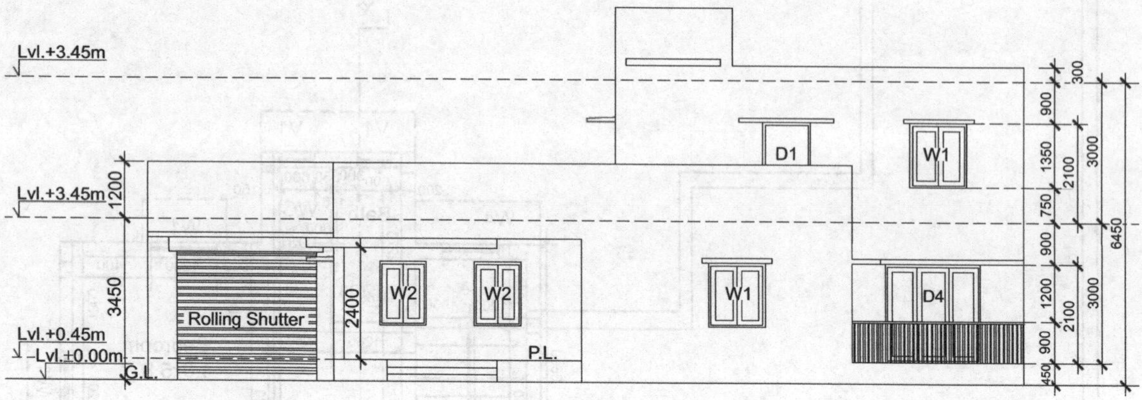

Fig. 4.11: Front Elevation

4.2.4 Section

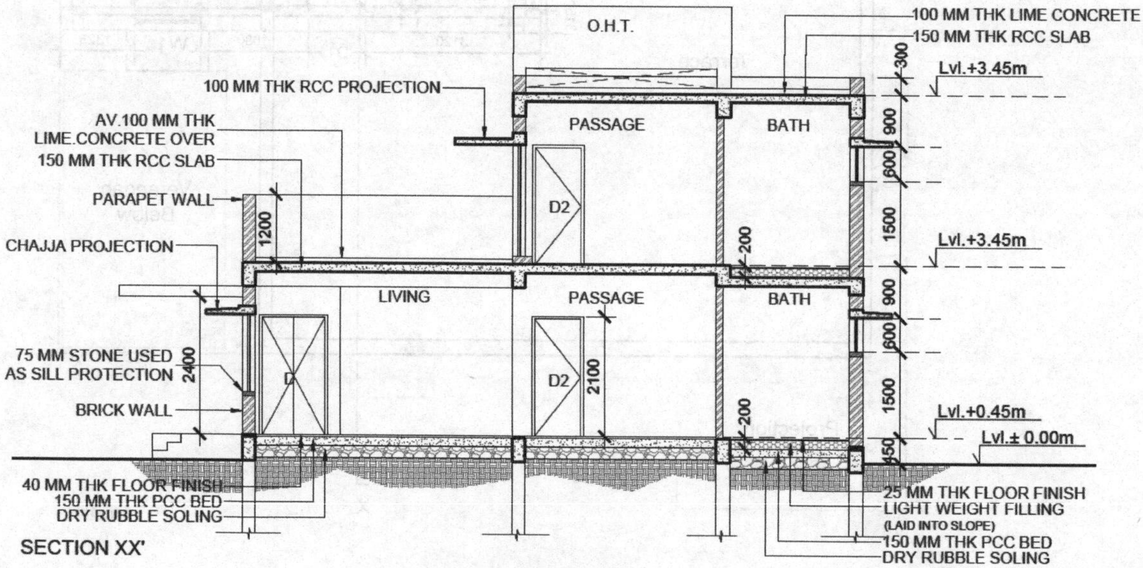

Fig. 4.12: Section XX

4.2.5 Working Drawing of Staircase

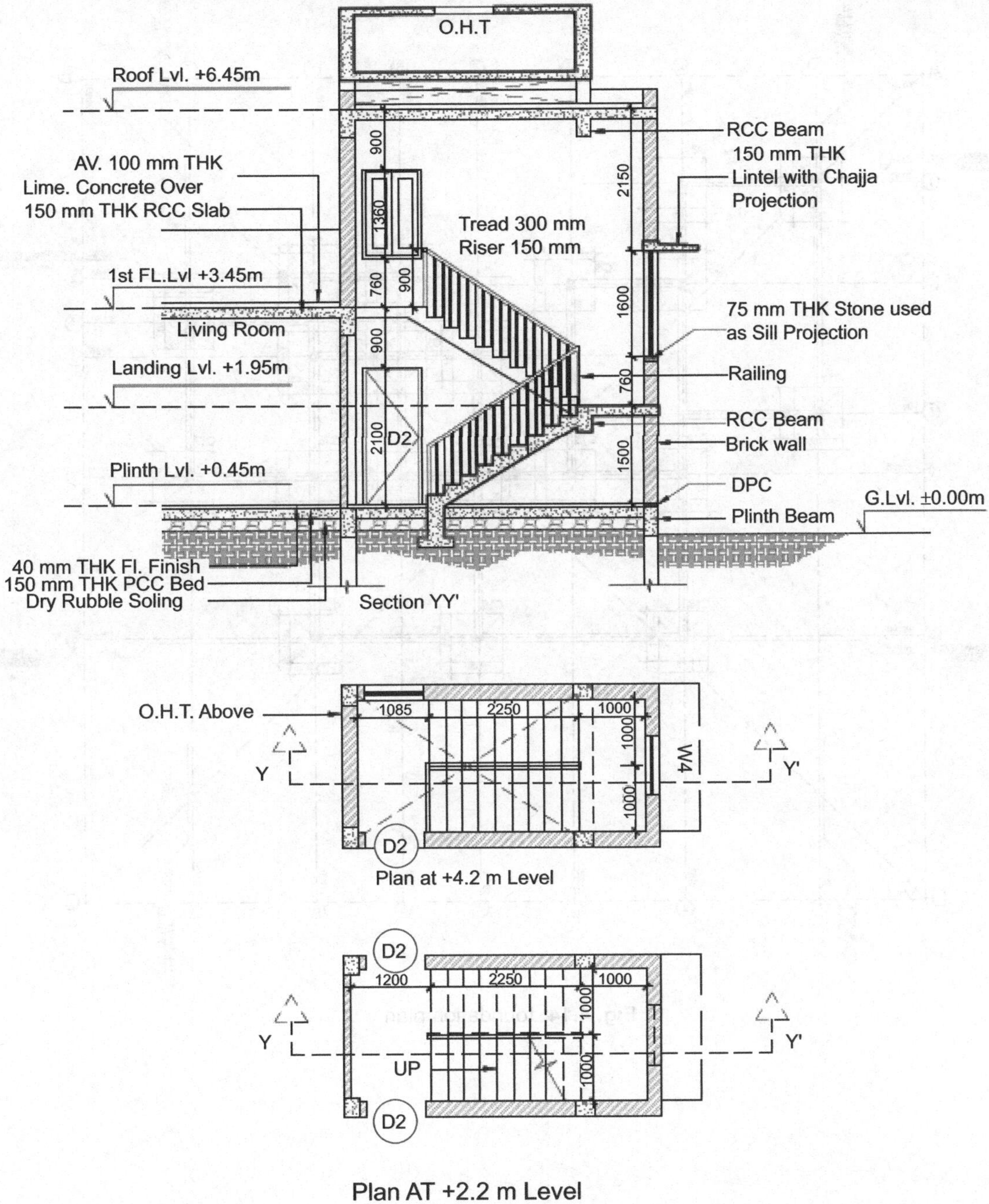

Fig. 4.13: Plan

4.4 FOUNDATION PLAN OF FRAME STRUCTURE.

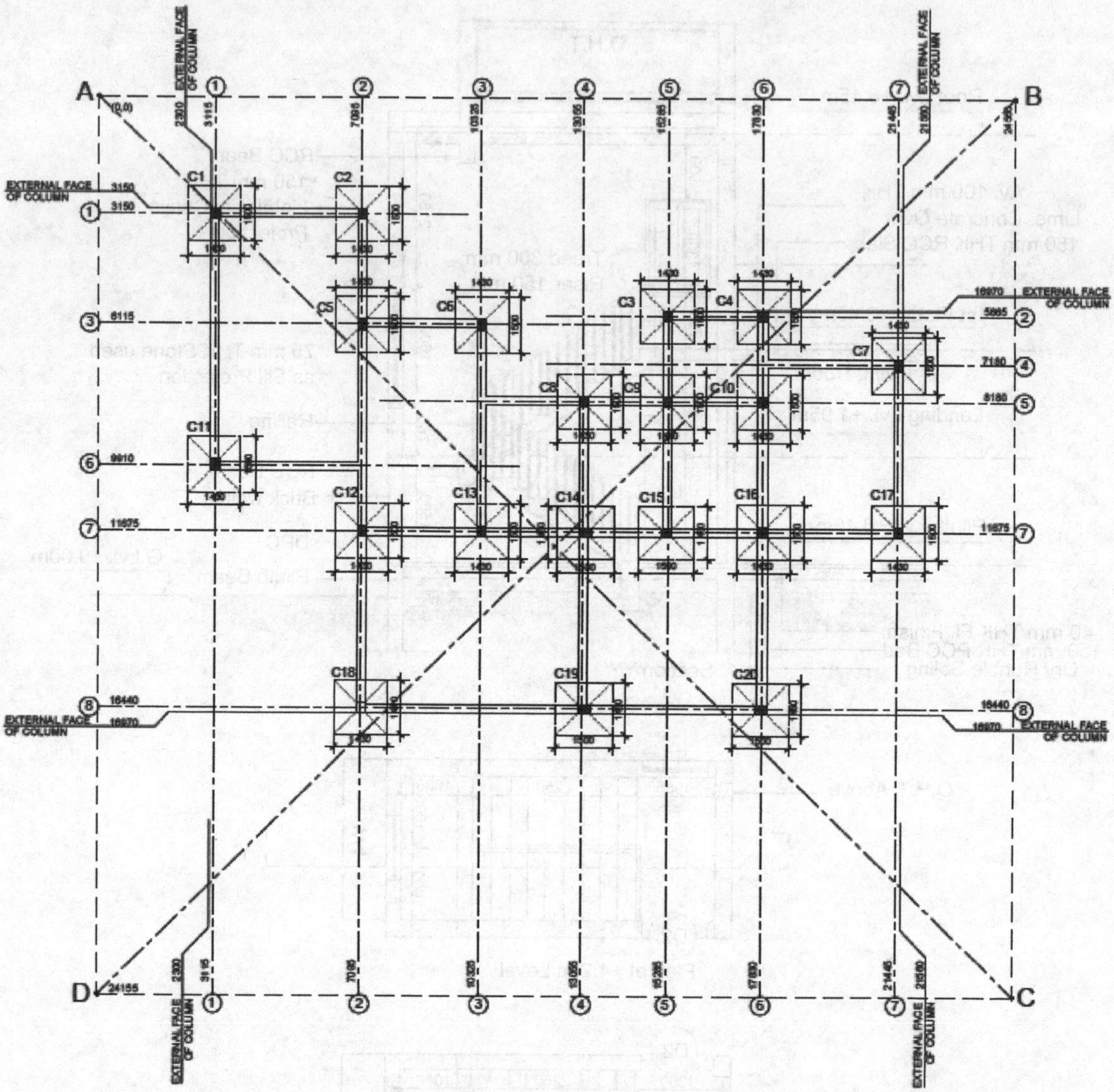

Fig. 4.14: foundation plan

4.5.1 Details of R.C.C Footings

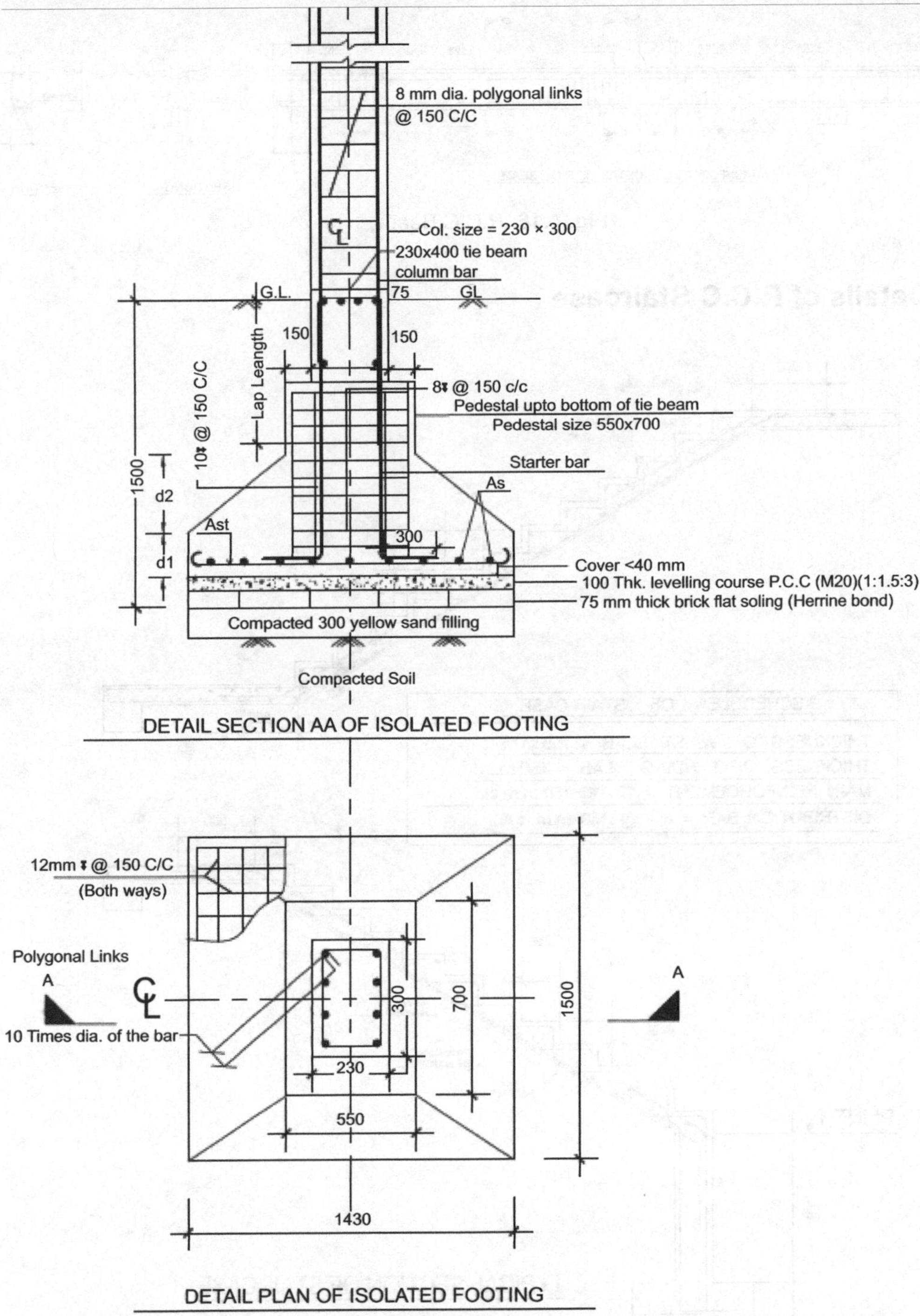

Fig. 4.15: R.C.C Column Footing

4.5.2 Details of R.C.C Beam

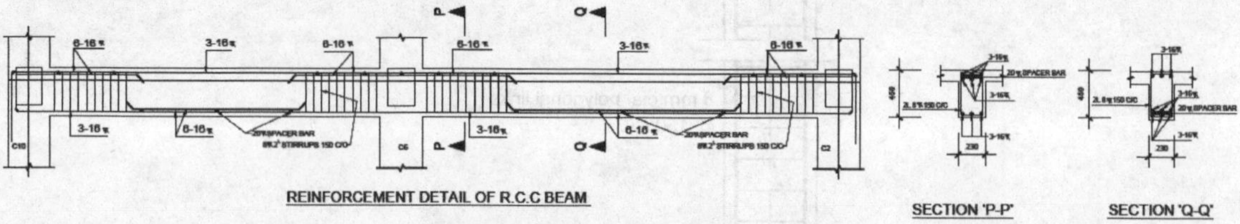

Fig. 4.16: R.C.C Beam

4.5.3 Details of R.C.C Staircase

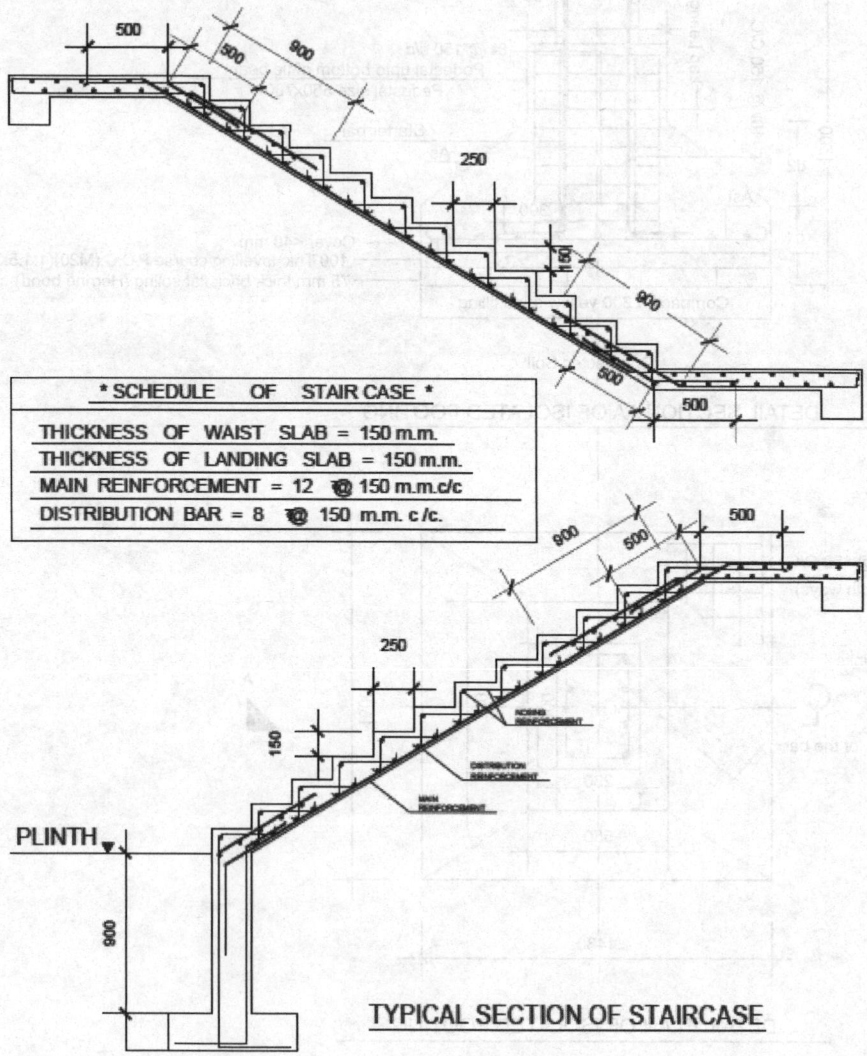

* SCHEDULE OF STAIR CASE *
THICKNESS OF WAIST SLAB = 150 m.m.
THICKNESS OF LANDING SLAB = 150 m.m.
MAIN REINFORCEMENT = 12 @ 150 m.m.c/c
DISTRIBUTION BAR = 8 @ 150 m.m. c/c.

Fig. 4.17: R.C.C Staircase

4.5.3 Details of R.C.C Slab

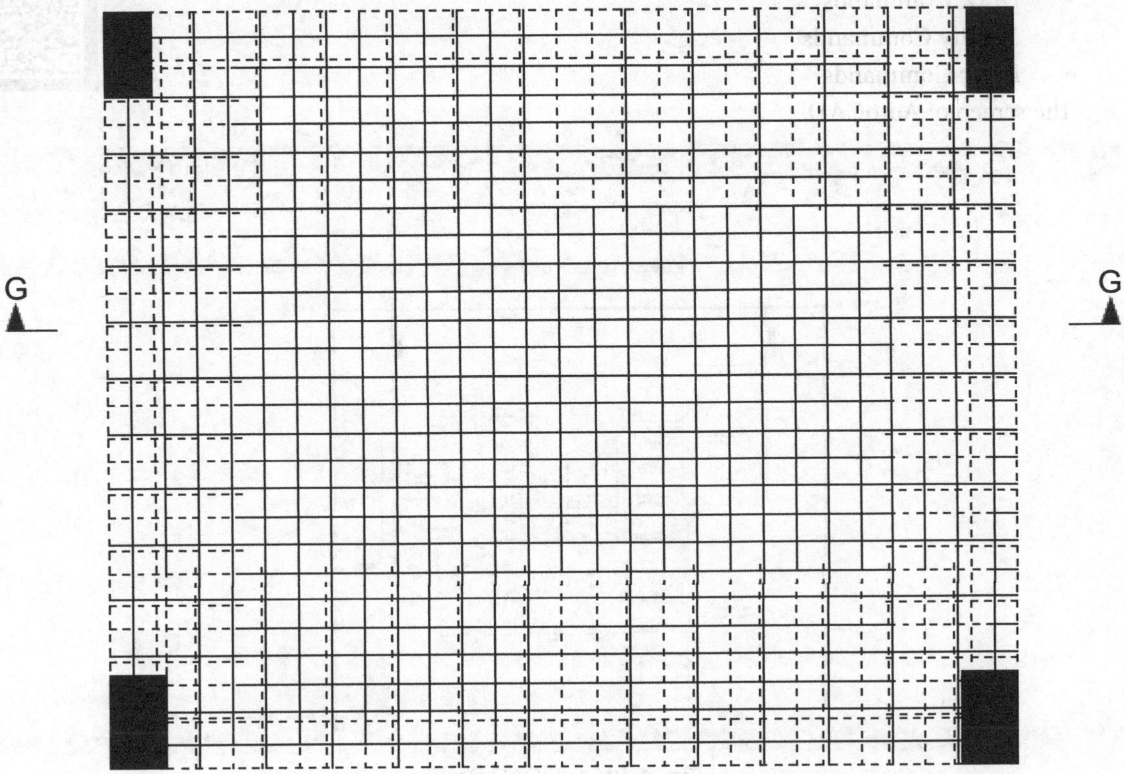

Detl. For Reinforcement Slab

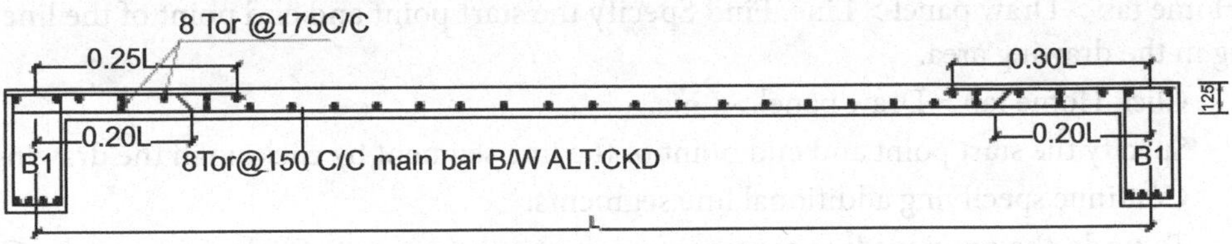

Section AT - G G

Fig. 4.18: R.C.C SLAB

4.6 DRAWING WITH CAD

CAD is Computer Aided Drafting and called as **Computer Aided Design**. It is tool for digital drafting. It is a very efficient and effective tool. First the designer needs to put the idea on the paper and then for the drafting purpose this tool is very useful. In this drawing, modifying, copying, erasing, making different layers, etc. is easy and fast. To preserve the drawing for long, it is easy to keep in computer and work on it later also. It can be used to make 2D as well as 3D drawings.

Certain commands which are very useful to start the drawing on the computer are

- Draw Commands
- Modify Commands
- Layer Commands

The screen of AutoCAD

Scan Me

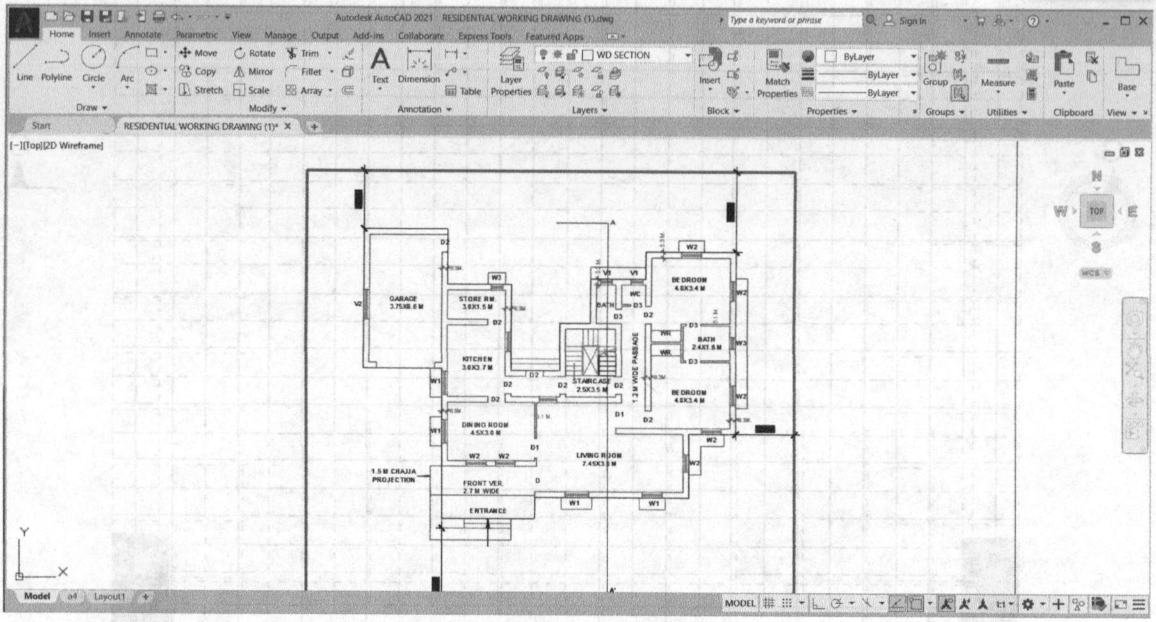

Fig. 4.19: AutoCAD Screen

4.6.1 How To Draw Lines

Click Home tab > Draw panel > Line. Find Specify the start point and end point of the line segment by clicking in the drawing area.

1. Click Home tab > Draw panel > Line.
2. Specify the start point and end point of the line segment by clicking in the drawing area.
3. Continue specifying additional line segments.

   To undo the previous line segment, enter **u** at the prompt. Click Undo on the Quick Access toolbar to cancel the entire series of line segments:

Fig. 4.20: Undo Command

4. Press Enter or Esc when done or enter **c** to close a series of line segments.

4.6.1.1 Using Specific Coordinates

Click Home tab > Draw panel > Line.

1. Type the coordinate value for the first point by typing the X value, a comma, then the Y value, for example 2.85,5.35.
2. Press the Spacebar or Enter.

Note: When dynamic input is on, relative coordinates are the default. When dynamic input is off, absolute coordinates are the default. Press F12 to turn dynamic input on or off.

4.6.1.2 At a Specific Angle

1. Click Home tab > Draw panel > Line.
2. Specify the start point.
3. Do one of the following to specify the angle:
    - Enter the left angle bracket (<) and the angle, for example <45, and move the cursor to indicate the direction.
    - Enter polar coordinates, for example 2.5<45.
    - Press F8 to turn on Ortho to lock the angle to horizontal and vertical directions. You can also press Shift when specifying the next point for a horizontal or vertical direction.
    - Press F10 to turn on polar tracking. You might need to use the DSETTINGS command, Polar Tracking tab to specify additional polar angles and choose the option to track all polar angle settings.
    - Move the cursor to indicate an approximate angle.
4. Do one of the following to specify the length:
    - Click a point to specify the endpoint with or without using object snaps. Trim or extend the resulting line as needed.
    - Enter the length of the line, for example 2.5.
5. Press the Spacebar or Enter.

4.6.1.3 At a Specific Angle to Another Line

Temporarily aligning the UCS icon with an existing line makes it easy to draw a line at a specific angle to another line.

1. At the command prompt, enter UCS.
2. Enter **ob** for Object and select the existing line.
   The UCS origin (0,0,0) is redefined.
3. Click Home tab > Draw panel > Line.
4. Specify the start point.
5. Do one of the following to specify the angle:
    - Enter the left angle bracket (<) and the angle, for example <45, and move the cursor to indicate the direction.
    - Move the cursor to indicate the approximate angle.
6. Specify the second point.
7. Press the Spacebar or Enter.

8. At the command prompt, enter UCS.
9. Enter **p** for Previous to reset the UCS origin.

4.6.1.4 To Draw Polylines

Draw polylines with straight or curved segments. Define the width for the polyline segments and taper the width across the segment. Draw polygons specifying the number of sides and size.

Draw a Polyline with Straight Segments

Click Home tab>Draw panel>Polyline.

1. Specify the first point of the polyline.
2. Specify the endpoint of the first segment.
3. Continue specifying segment endpoints as needed.
4. Press Enter to end or enter **c** to close the polyline.

Note: To start a polyline at the endpoint of the last polyline drawn, start the command again and press Enter at the Specify Start Point prompt.

4.6.1.5 To Draw a Multiline

To change the scale of the multiline, enter s and enter a new scale. Now **draw** the multiline. Specify the starting point.

1. At the command prompt, enter MLINE.
2. At the Command prompt, enter **st** to select a style.
3. To list available styles, enter the style name or enter?
4. To justify the multiline, enter **j** and select top, zero, or bottom justification.
5. To change the scale of the multiline, enter **s** and enter a new scale.
   Now draw the multiline.
6. Specify the starting point.
7. Specify a second point.
8. Specify additional points or press Enter.
   If you specify three or more points, you can enter **c** to close the multiline

4.6.1.6 Draw a Polyline with Straight and Curved Segments

1. Click Home tab>Draw panel>Polyline.
2. Specify the first point of the polyline.
3. Specify the endpoint of the first segment.
4. Switch to Arc mode by entering **a** (Arc) at the Command prompt.
5. Return to Line mode by entering **L** (Line).
6. Specify additional segments as needed.
7. Press Enter to end or enter **c** to close the polyline.

4.6.1.7 To Draw a Rectangle

1. Click Home tab>Draw panel>Rectangle.
2. Specify the first corner of the rectangle.

3. Specify the diagonal corner of the rectangle.

4.6.1.8 Draw a Polygon

1. Click Home tab > Draw panel > Polygon.
2. Enter the number of sides.
3. Specify the center of the polygon.
4. Do one of the following:

   Enter **i** to specify a polygon inscribed within a circle.

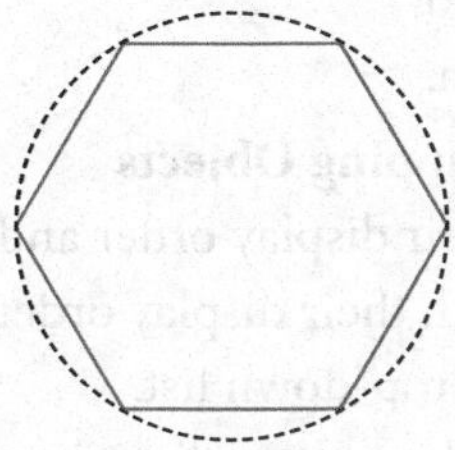

Fig. 4.21: Circle Command

   Enter **c** to specify a polygon circumscribed about a circle.

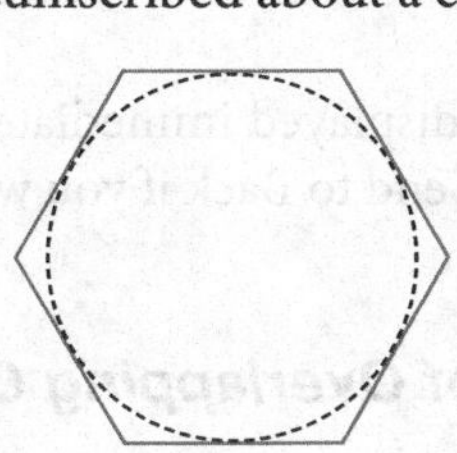

Fig. 4.22: Circle Command

5. Enter the radius length.

4.6.1.9 Draw a Boundary Polyline

1. Click Home tab>Draw panel>Boundary.
2. In the Boundary Creation dialog box, Object Type list, select Polyline.
3. Click Pick Points. Specify points within each area to form a boundary polyline for each.

   Note: Each area must be enclosed.
4. Press Enter to create the boundary polylines and end the command.

   Note: The boundary polyline overlaps the objects used to create it.

4.6.1.10 To Draw Freehand Sketches

At the Command prompt, enter sketch. Press Enter again to accept the last saved type, increment, and tolerance

1. At the Command prompt, enter **sketch.**
2. Press Enter again to accept the last saved type, increment, and tolerance values.
3. Move cursor in the drawing area to begin sketching.

As you move the pointing device, freehand line segments of the specified length are drawn. During the command, freehand lines are displayed in a different color.

4. Click to suspend sketching.
   You can click a new start point to resume drawing from the new cursor location.
5. Press Enter to complete the sketch.

(a) An alternate method

1. At the Command prompt, enter **sketch**.
2. Click and hold to begin sketching and move your cursor. Release to suspend sketching.
3. Repeat the previous step as needed.
4. Press Enter to complete the sketch.

(b) To Control the Draw Order of Overlapping Objects

The **draw** order of objects includes both their display order and plotting order.

The draw order of objects includes both their display order and plotting order.

1. Click Home tab > Modify panel drop-down list.
2. Click Draw Order, and from the drop-down list, choose one of the options.
3. Select the objects that you want to modify, and press Enter.
4. For the Bring Above Objects and Send Under Objects options, select the reference objects and press Enter.

Note: The objects being modified are displayed immediately above or under the objects you select as reference objects. Use Bring to Front or Send to Back if you want the objects displayed above or below all objects.

4.6.1.11 About the Draw Order of Overlapping Objects

One can change the **draw** order, which is the display and plotting order, of specified objects. One can change the draw order, which is the display and plotting order, of specified objects. Can control overlapping objects that cover an area, such as objects with thick lineweights, wide polylines, hatches and fills, annotation, and images. For example, the following map includes several different types of overlapping objects. Obviously, the order in which these objects display and plot is important.

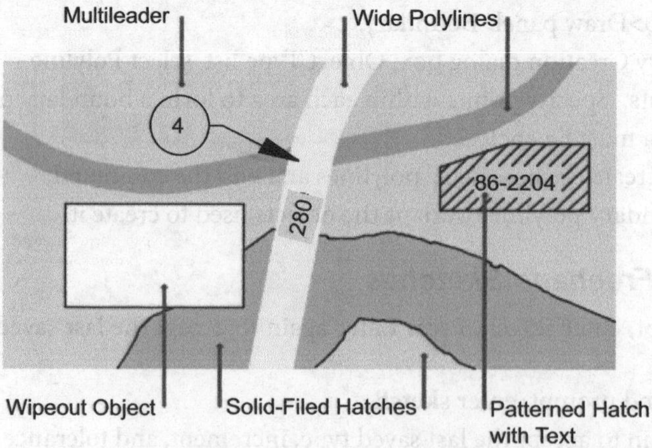

Fig. 4.23: Overlapping Objects

4.6.1.12 About Arcs

Hold down the Ctrl key as you drag to **draw** in a clockwise direction.

Create arcs by specifying various combinations of center, endpoint, start point, radius, angle, chord length, and direction values.

Arcs are drawn in a counterclockwise direction by default. Hold down the Ctrl key as you drag to draw in a clockwise direction.

4.6.1.13 Draw Arcs by Specifying Three Points

Click Home tab > Draw panel > Arc drop-down > 3-Point.

(a) Draw Arcs by Specifying Start, Center, End

You can create an arc using a start point, center, and a third point that determines the endpoint.

The distance between the start point and the center determines the radius. The endpoint is determined by a line from the center that passes through the third point.

Using different options, you can specify either the start point first or the center point first.

Click Home tab > Draw panel > Arc drop-down > Start, Center, End.

Click Home tab > Draw panel > Arc drop-down > Center, Start, End.

(b) Draw Arcs by Specifying Start, Center, Angle

You can create an arc using a start point, center, and an included angle.

The distance between the start point and the center determines the radius. The other end of the arc is determined by specifying an included angle that uses the center of the arc as the vertex.

Using different options, you can specify either the start point first or the center point first.

The included angle determines the endpoint of the arc. Use the Start, End, Angle method when you know both endpoints but cannot snap to a center point.

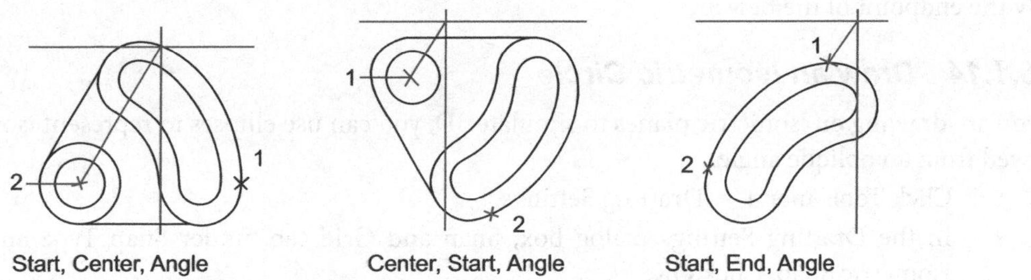

Fig. 4.24: Different ways of Arcs

Click Home tab > Draw panel > Arc drop-down > Start, Center, Angle.

Click Home tab > Draw panel > Arc drop-down > Center, Start, Angle.

Click Home tab > Draw panel > Arc drop-down > Start, End, Angle.

(c) Draw Arcs by Specifying Start, Center, Length

One can create an arc using a start point, center, and the length of a chord.

The distance between the start points and the center determines the radius. The other end of the arc is determined by specifying the length of a chord between the start point and the endpoint of the arc.

Using different options, you can specify either the start point first or the center point first.

The length of the chord of the arc determines the included angle.

Click Home tab > Draw panel > Arc drop-down > Start, Center, Length.

Click Home tab > Draw panel > Arc drop-down > Center, Start, Length.

(d) Draw Arcs by Specifying Start, End, Angle

You can create an arc using a start point, endpoint, and an included angle.

The included angle between the endpoints of the arc determines the center and the radius of the arc.

Click Home tab > Draw panel > Arc drop-down > Start, End, Angle.

(e) Draw Arcs by Specifying Start, End, Direction

You can create an arc using a start point, endpoint, and a tangent direction at the start point.

The tangent direction can be specified either by locating a point on the desired tangent line, or by entering an angle. You can determine which endpoint controls the tangent by changing the order in which you specify the two endpoints.

(f) Draw Arcs by Specifying Start, End, Radius

You can create an arc using a start point, endpoint, and a radius.

The direction of the bulge of the arc is determined by the order in which you specify its endpoints. You can specify the radius either by entering it or by specifying a point at the desired radius distance. Click Home tab > Draw panel > Arc drop-down > Start, End, Radius.

Click Home tab > Draw panel > Arc drop-down > Start, End, Direction.

(g) Draw Contiguous Tangent Arcs and Lines

Immediately after you create an arc, you can start a line that is tangent to the arc at an endpoint. You need to specify only the line length.

Immediately after you create a line or an arc, you can start an arc that is tangent at an endpoint by starting the ARC command and pressing Enter at the Specify Start Point prompt. You need to specify only the endpoint of the new arc.

4.6.1.14 Draw an Isometric Circle

If you are drawing on isometric planes to simulate 3D, you can use ellipses to represent isometric circles viewed from an oblique angle.

1. Click Tools menu ➤ Drafting Settings.
2. In the Drafting Settings dialog box, Snap and Grid tab, under Snap Type and Style, click Isometric Snap. Click OK.
3. Click Home tab ➤ Draw panel ➤ Ellipse drop-down ➤ Axis, End.
4. Enter **i** (Isocircle).
5. Specify the center of the circle.
6. Specify the radius or diameter of the circle.

4.6.1.15 Draw a True Ellipse Using Endpoints and Distance

1. Click Home tab ➤ Draw panel ➤ Ellipse drop-down ➤ Axis, End.
2. Specify the first endpoint of the first axis (1).
3. Specify the second endpoint of the first axis (2).

4. Drag the pointing device away from the midpoint and click to specify a distance (3) for half the length of the second axis.

4.6.1.16 Draw an Elliptical Arc Using Start and End Angles

Click Home tab ➤ Draw panel ➤ Ellipse drop-down ➤ Elliptical Arc.

1. Specify endpoints for the first axis (1 and 2).
2. Specify a distance to define half the length of the second axis (3).
3. Specify the start angle (4).
4. Specify the end angle (5).

4.6.2 Modify Commands

Commands for Modifying Dimension Geometry

4.6.2.1 DIMEDIT

- Edits dimension text and extension lines.
- Rotates, modifies, or restores dimension text. Changes the oblique angle of extension lines. The companion command that moves text and the dimension line is DIMTEDIT.
- The following prompts are displayed.

(a) Home

- Moves rotated dimension text back to its default position.

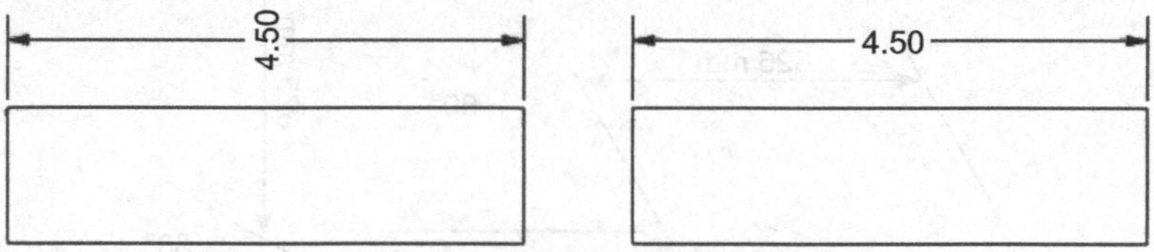

Fig. 4.25: Modify Dimensions

- The selected dimension text is returned to its default position and rotation as specified in its dimension style.

(b) New

- Changes dimension text using the In-Place Text Editor.

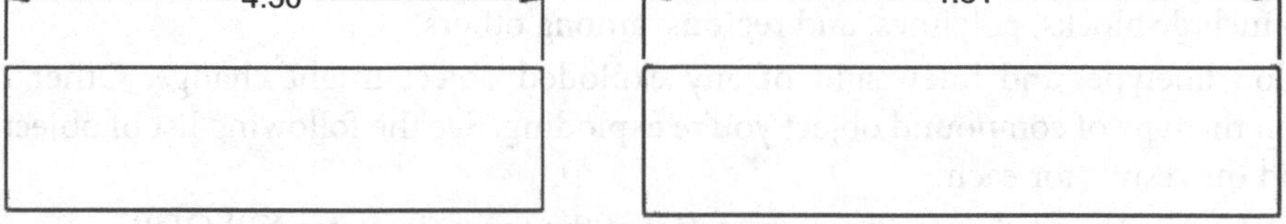

Fig. 4.26: Modify Text

The generated measurement is represented with angle brackets (< >). Use control codes and Unicode character strings to enter special characters or symbols. See Control Codes and Special Characters.

To edit or replace the generated measurement, delete the angle brackets, enter the new dimension text, and then choose OK. If alternate units are not turned on in the dimension style, you can display them by entering square brackets ([ ]).

(c) Rotate

Rotates dimension text. This option is similar to the Angle option of DIMTEDIT.

Entering 0 places the text in its default orientation, which is determined by the vertical and horizontal text settings on the Text tab of the New, Modify, and Override Dimension Style dialog boxes. See DIMSTYLE. The DIMTIH and DIMTOH system variables control this orientation.

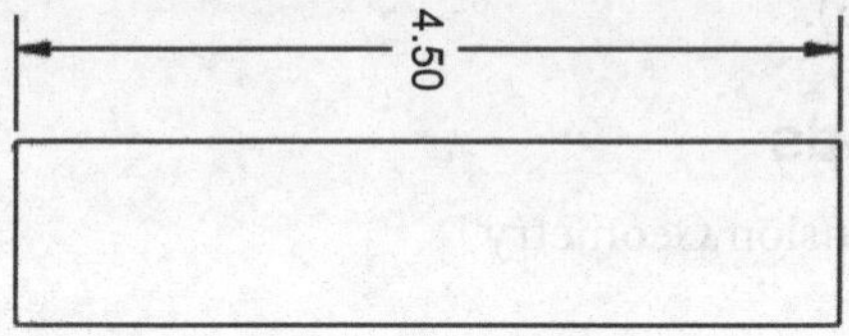

Fig. 4.27: Rotate Text

(d) Oblique

The Oblique option is useful when extension lines conflict with other features of the drawing. The oblique angle is measured from the X axis of the UCS.

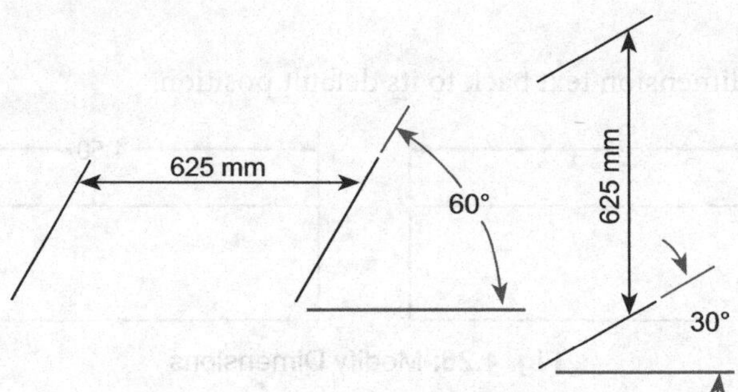

Fig. 4.28: Dimension at an angle

(e) Explode

Breaks a compound object into its component objects.

Explodes a compound object when you want to modify its components separately. Objects that can be exploded include blocks, polylines, and regions, among others.

The color, linetype, and lineweight of any exploded object might change. Other results differ depending on the type of compound object you're exploding. See the following list of objects that can be exploded and the results for each.

To explode objects and change their properties at the same time, use XPLODE.

(f) Stretch

Stretches objects crossed by a selection window or polygon.

Objects that are partially enclosed by a crossing window are stretched. Objects that are completely enclosed within the crossing window, or that are selected individually, are moved rather than stretched. Some types of objects such as circles, ellipses, and blocks, cannot be stretched.

(g) Stretch

Moves only the vertices and endpoints that lie inside the crossing selection, leaving those outside unchanged. STRETCH does not modify 3D solids, polyline width, tangent, or curve-fitting information.

(h) Xplode

Breaks a compound object into its component objects, with specified properties for the resulting objects. It also reports how many objects were selected and, of those, how many objects cannot be exploded.

The following prompt is displayed for each object.

All: Sets the color, linetype, lineweight, and layer of the component objects after you explode them. The prompts associated with the Color, Linetype, Lineweight, and Layer options are displayed.

Color: Sets the color of the objects after you explode them.

Enter **bylayer** to inherit the color of the exploded object›s layer.

Enter **byblock** to inherit the color of the exploded object.

Enter **t** for a true color to be used for the selected object.

Enter **co** for a color from a loaded color book to be used for the selected object.

Layer: Sets the layer of the component objects after you explode them. The default option is to inherit the current layer rather than the layer of the exploded object.

LType: Sets the linetype of the component objects after you explode them.

Enter **bylayer** to inherit the linetype of the exploded object's layer.

Enter **byblock** to inherit the linetype of the exploded object.

LWeight: Sets the lineweight of the component objects after you explode them.

Explode: Breaks a compound object into its component objects exactly as the EXPLODE command does.

4.6.2.2 TEXTEDIT

Edits a selected multiline or single-line text object, or the text in a dimension object. Displays the in-place text editor, and accepts your changes to the selected multiline text, single-line text, or dimension object.

The following prompts are displayed.

Select an annotation object: Specifies the text, multiline text, or dimension object that you want to edit.

Undo: Undoes the last change made to the text object.

Mode: Controls whether the command repeats automatically. (TEXTEDITMODE system variable)

Single: Modifies the selected text object once and ends the command.

Multiple: Enters the Multiple mode, which allows you to edit multiple text objects within the command duration.

4.6.2.3 To Modify Multilines

Select the multiline for the background. The intersection is modified. You can continue selecting intersecting multilines to modify, or press Enter to end the command. Multiline objects are composed of 1 to 16 parallel lines, called *elements.*

(a) Edit a Multiline Style

Multiline styles control the number of line elements in a multiline and the color, linetype, lineweight, and offset of each element. You can also modify the display of joints, end caps, and background fill.

1. At the Command prompt, enter MLSTYLE.
2. In the Multiline Styles dialog box, select the style name from the list. Click Modify.
3. Click Element Properties.
4. In the Modify Multiline Styles dialog box, change the settings as needed.
5. Click OK.
6. In the Multiline Styles dialog box, click Save to save the changes to the style in the MLN file.

(b) Delete a Vertex from a Multiline

1. At the Command prompt, enter MLEDIT.
2. In the Multiline Edit Tools dialog box, select Delete Vertex.
3. In the drawing, specify the vertex to delete. Press Enter.

(c) Create a Closed Cross Intersection

Multilines can intersect in a cross or a **T** shape, and the crosses or **T** shapes can be closed, open, or merged.

1. At the Command prompt, enter MLEDIT.
2. In the Multiline Edit Tools dialog box, select Closed Cross.
3. Select the multiline for the foreground.
4. Select the multiline for the background.

The intersection is modified. You can continue selecting intersecting multilines to modify, or press Enter to end the command.

(d) To Modify the Properties of a Hatch Object

Right-click to access the Hatch Edit and other **commands.**

There are a few different tools to help you make property changes to existing hatch objects. Once you've selected a hatch object, you can:

- Use the Hatch Editor ribbon controls.
- Hover over the hatch control grip to display a dynamic menu that lets you quickly change the pattern origin point, angle, and scale.

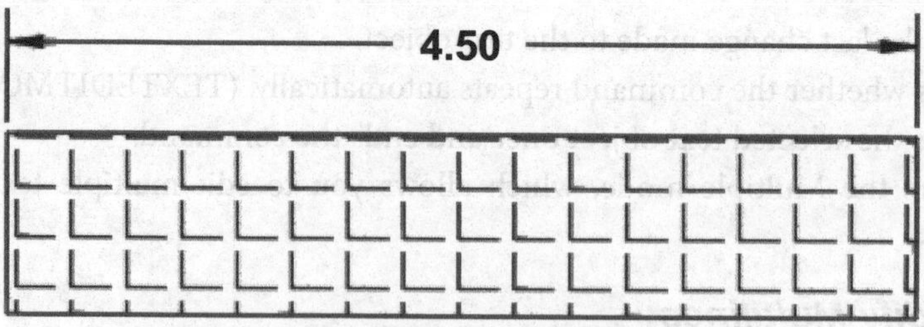

Fig. 4.29: Hatch an object

- Use the Properties palette.
- Right-click to access the Hatch Edit and other commands.

4.6.3 Layer Commands

These are different layers of the sheets on which one can work. On every layer of the sheet one can change properties like colour, linetype, lineweights etc. if any object has different colour is set to BYELAYER, that object will be seen in that colour regardless of the colour assign to it.

Manages layers and layer properties. The Layer Properties Manager is displayed.

Enter -LAYER at the **Command prompt**, options are displayed.

Displays a list of the currently defined layers, showing their names, states, color numbers, linetypes, lineweights, and whether they are externally dependent layers.

Make: Creates a layer and makes it current. New objects are drawn on the current layer.

If no layer exists for the name you enter, a new layer with that name is created. The new layer is on and assumes the following properties by default: color number 7, the CONTINUOUS linetype, and a lineweight of DEFAULT.

If the layer exists but is turned off, it is turned on.

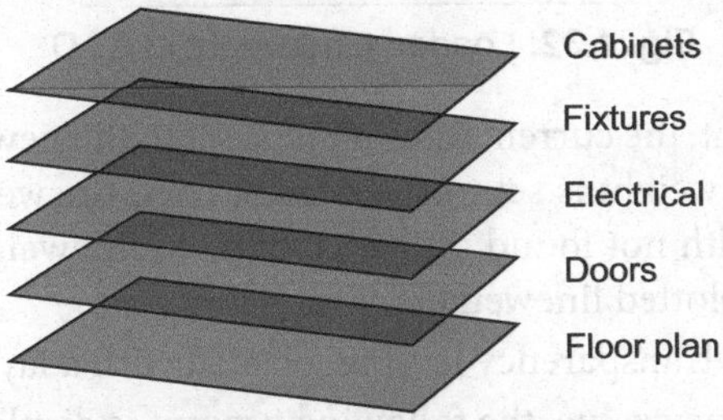

Fig. 4.30: Make LAYERS of Different items

Set: Specifies a new current layer but does not create the layer if it does not already exist. If the layer exists but is turned off, it is turned on and made current. A frozen layer cannot be made current.

New: Creates layers. You can create two or more layers by entering names separated by commas.

Rename: Renames an existing layer.

On: Makes selected layers visible and available for plotting.

Off: Makes selected layers invisible and excludes them from plotting.

Color: Changes the color associated with a layer. Enter a color name or a number from 1 through 255.

Ltype: Changes the linetype associated with a layer.

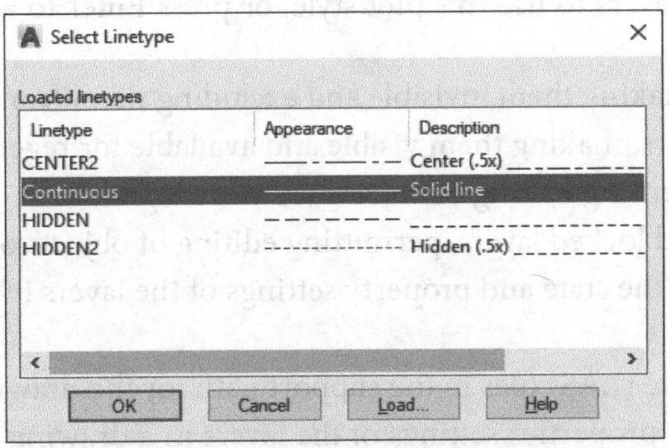

Fig. 4.31: Loaded Linetype in CAD

Lweight: Changes the lineweight associated with a layer.

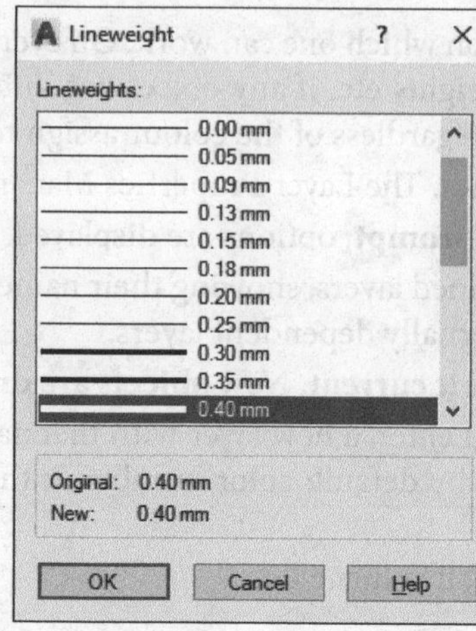

Fig. 4.32: Loaded Lineweight in CAD

If you enter a valid lineweight, the current lineweight is set to the new value. If you enter a lineweight that is not valid, the current lineweight is set to the nearest fixed lineweight value. If you would like to plot an object with a custom width not found in the list of fixed lineweight values, you can use the Plot Style Table Editor to customize plotted lineweights.

Transparency: Changes the transparency level associated with a layer. Enter a value from 0 to 90.

After specifying a transparency value, the following prompt is displayed:

Enter name list of layer(s) for transparency specified value% <0>: Enter the names of the layers to which to apply this transparency level, or press Enter to apply it to the current layer only.

Material: Attaches a material to a layer. The material must be available in the drawing before it can be assigned to a layer.

Plot: Controls whether visible layers are plotted. If a layer is set to plot but is currently frozen or turned off, the layer is not plotted.

Pstyle: Sets the plot style assigned to a layer. The Pstyle option is available only when you are using named plot styles.

If you select a plot style other than NORMAL, the following prompt is displayed:

Enter name list of layers(s) for plot style current <current>:

Enter the names of the layers to use this plot style, or press Enter to apply the style to the current layer only

Freeze: Freezes layers, making them invisible and excluding them from regeneration and plotting.

Thaw: Thaws frozen layers, making them visible and available for regeneration and plotting.

Lock: Locks layers, preventing editing of objects on those layers.

Unlock: Unlocks selected locked layers, permitting editing of objects on those layers.

State: Saves and restores the state and property settings of the layers in a drawing.

List Named Layer States

Lists the named layer state (LAS) files in the support path for the drawing.

Save: Saves the state and properties settings of the layers in a drawing under a specified layer state name. When saving a layer state, you specify which layer settings are affected when the layer state is later restored.

Restore: Restores the state and property settings of all layers to previously saved settings. Restores only those layer state and property settings that were selected when the layer state was saved.

Edit: Changes the saved layer settings for a specified layer state. When the layer state is restored, the specified settings are used.

Name: Changes the name of a saved layer state.

Delete: Removes a saved layer state.

Import: Loads a previously exported layer state (LAS) file, or layers states from a file (DWG, DWS, DWT) into the current drawing. Additional layers may be created because of importing a layer state file.

Export: Saves a selected named layer state to a layer state (LAS) file.

(a) Layer Settings Dialog Box (Layer Properties Manager)

Find: Layer Settings Controls when notification occurs for New layers, layer behavior when some layer are isolated, whether layer filters are applied to the Layers toolbar, and the background color of viewport overrides in the Layer Properties Manager.

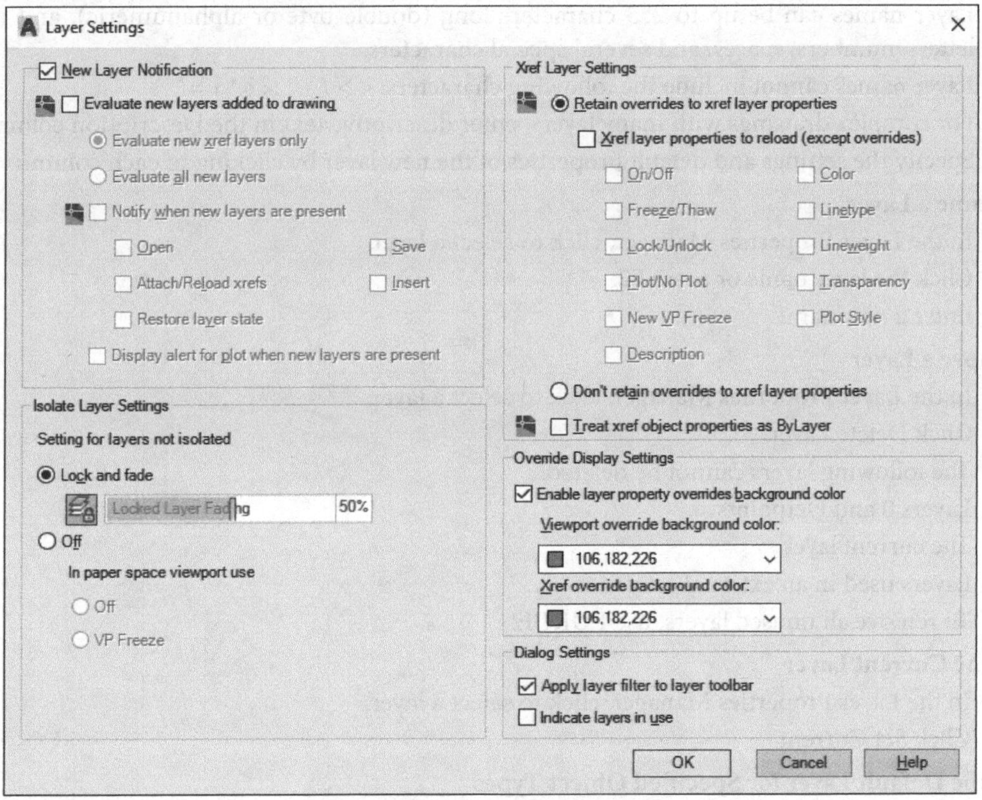

Fig. 4.33: Layer Setting in CAD

- Controls when notification occurs for new layers, layer behavior when some layer are isolated, whether layer filters are applied to the Layers toolbar, and the background color of viewport overrides in the Layer Properties Manager.

(b) To Work with Layers

- Click Set Current. Set the Default Layer for Specified Object Types At the Command prompt, enter the system variable that controls the default layer for a specified object type: Center marks and centerlines: CENTERLAYER system variable Hatches and fills: HPLAYER system variable Dimensions: DIMLAYER system variable Mtext and text: TEXTLAYER system variable Xrefs: XREFLAYER system variable Enter the name of the layer to which subsequently created object types are to be assigned.
- Create, rename, and remove layers. Set the current layer. Change layer properties such as color, linetype, lineweight, and transparency.
- Use the Layer Properties Manager to work with layers. Click Home tab ➤ Layers panel ➤ Layer Properties.

(c) Create a Layer

- In the Layer Properties Manager, click New Layer.
- A layer name is added to the layer list.
- Enter a new layer name by typing over the highlighted layer name.
- Layer names can be up to 255 characters long (double-byte or alphanumeric), and include letters, numbers, spaces, and several special characters.
- Layer names cannot include the following characters: < > / \ " : ; ? * | = '
- For complex drawings with many layers, enter descriptive text in the Description column.
- Specify the settings and default properties of the new layer by clicking in each column.

(d) Rename a Layer

- In the Layer Properties Manager, click to select a layer.
- Click the layer name or press F2.
- Enter a new name.

(e) Remove a Layer

- In the Layer Properties Manager, click to select a layer.
- Click Delete Layer.
- The following layers cannot be deleted:
- Layers 0 and Defpoints
- The current layer
- Layers used in an external reference
- To remove all unused layers, use PURGE.

(f) Set the Current Layer

- In the Layer Properties Manager, click to select a layer.
- Click Set Current.

(g) Set the Default Layer for Specified Object Types

- At the Command prompt, enter the system variable that controls the default layer for a specified object type:
- Center marks and centerlines: CENTERLAYER system variable
- Hatches and fills: HPLAYER system variable
- Dimensions: DIMLAYER system variable

- Mtext and text: TEXTLAYER system variable (added in the 2020.1 update)
- Xrefs: XREFLAYER system variable
- Enter the name of the layer to which subsequently created object types are to be assigned.

(h) Change the Properties Assigned to Layers

- Want to change multiple layers, use one of the following methods in the Layer Properties Manager:
- Press and hold Ctrl and choose several layer names.
- Press and hold Shift and choose the first and last layers in a range.
- Right-click and click Show Filters in Layer List. Choose a layer filter from the list of layers.
- Click the current setting in the column that you want to change.
- The dialog box for that property displays.
- Choose the setting that you want to use.
- When changing layer properties:
- If the linetype you want is not displayed, click Load and use one of the following methods:
    - In the Load or Reload Linetypes dialog box, choose the linetypes to load.
    - In the Load or Reload Linetypes dialog box, click File to open an additional linetype definition (LIN) file. Choose the linetypes to load and click OK.
- Lineweights are not displayed automatically. If you want to display or hide lineweights, click Show/Hide Lineweight on the status bar.
  If no change is visible, it's probably due to a combination of the thickness of the line compared to the display resolution of your monitor.
- Transparency is not displayed automatically. If you want to display or hide the transparency of objects, click Show/Hide Transparency on the status bar.

(i) View a List of Unused Layers

1. In the Layer Properties Manager, click Settings.
2. Check Indicate Layers in Use on the Layer Settings dialog box and click OK.
3. Click the Status column label to sort by status.

4.6.3.1 Layer Properties Manager

In LAYER command Designer can add, delete, and rename layers, change their properties, set property overrides in layout viewports, and add layer descriptions.

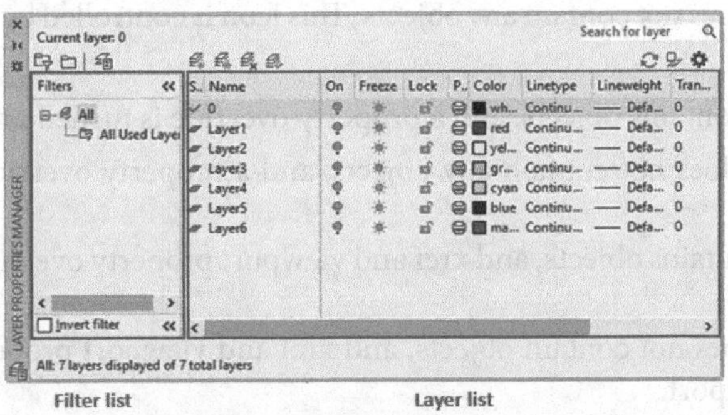

Fig. 4.34: Layer Properties in CAD

(a) New Layer 

Creates a layer with a default name that you can immediately change. The new layer inherits the properties of the currently selected layer in the layer list.

(b) New Layer Frozen VP In All Viewports 

Creates a layer and freezes it in all existing layout viewports. This button is accessible from the Model tab or layout tabs.

(c) Delete Layer 

Deletes selected layers. The following layers cannot be deleted:

- Layers 0 and Defpoints
- The current layer
- Layers used in a block definition
- Layers used in an external reference
- Layers in a partially opened drawing (Not applicable to AutoCAD LT.)

Note: Be careful about deleting layers if you are working on a drawing in a shared project or one based on a set of layer standards.

(d) Set Current 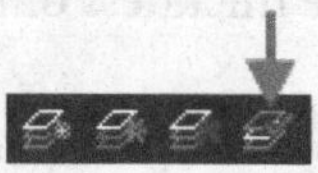

Sets the selected layer as the current layer. New objects are automatically created on the current layer. (CLAYER system variable)

(e) Layer List

Modify the layer properties using the layer list. Click the current setting to change the layer property for the selected layer or group of layers.

Column Order: Drag a column to a new location in the list to change the column order.

Status

- ✔ This layer is the current layer.
- This layer contains objects.
- This layer does not contain any objects. This icon is controlled by the SHOWLAYERUSAGE system variable.
- This layer contains objects, and a property override is turned on in a layout viewport.
- This layer does not contain any objects, and a property override is turned on in a layout viewport.
- This layer contains objects, and xref and viewport property overrides is turned on in a layout viewport.
- This layer does not contain objects, and xref and viewport property overrides is turned on in a layout viewport.
- This layer contains objects, and an xref property override is turned on.

- This layer does not contain objects, and an xref property override is turned on.

Name: Displays the name of the layer or filter. Press F2 to enter a new name.

On: Turns the selected layers on and off. When a layer is on, it is visible and available for plotting. When a layer is off, it is invisible and not plotted, even if the setting in the Plot column is turned on.

Freeze: Freezes the selected layers. You can freeze layers to improve performance and reduce regeneration time in complex drawings. Objects on frozen layers are not displayed, plotted, or regenerated.

Lock: Locks and unlocks the selected layers. Objects on a locked layer cannot be modified. Objects on locked layers appear faded and a small lock icon is displayed when you hover over the object.

Color ■ □: Displays the Select Color dialog box, where you can specify a color for the selected layers.

Linetype: Displays the Select Linetype dialog box, where you can specify a linetype for the selected layers.

Lineweight: Displays the Lineweight dialog box, where you can specify a lineweight for the selected layers.

Transparency: Displays the Transparency dialog box, where you can specify a transparency for the selected layers. Valid values are from 0 to 90. The higher the value, the more transparent the objects appear.

Plot Style: Displays the Select Plot Style dialog box where you can change the plot style for the selected layers.

Plot: Controls whether the selected layers are plotted. If you turn off plotting for a layer, the objects on that layer are still displayed. Layers that are off or frozen are not plotted, regardless of the Plot setting.

4.6.3.2 Select Layer States Dialog Box

LAYER (Command) Find List of Options The following option is available in this dialog box.

Displays a list of layer states to be imported into the current drawing.

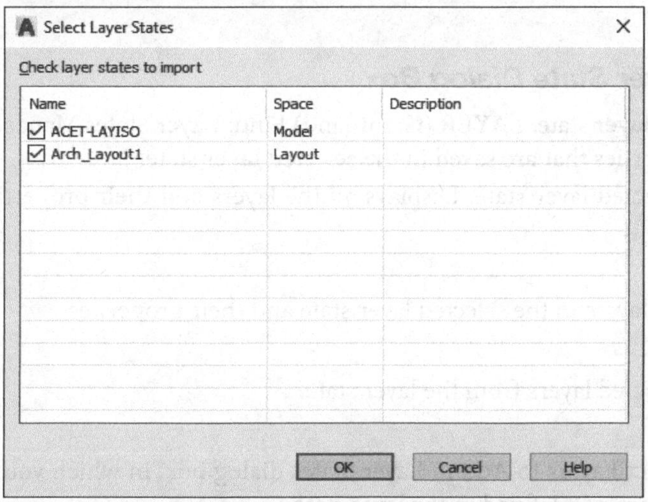

Fig. 4.35: Select Layer State to Import in CAD

Check Layer States to Import

Select the names of the layer states that you want to import.

The layer state names in the selected file (DWG, DWS, DWT) are compared to the layer state names in the current drawing. Duplicate layer state names are not check marked by default. When a duplicate layer state name is check marked, a prompt queries whether you want to overwrite the existing layer state.

4.6.3.3 Select Layers to Add to Layer State Dialog Box

LAYER (Command) Find Displays **layers** by name and includes descriptions that are not defined in the selected named **layer** state.

Specifies which layers should be included in the named layer state.

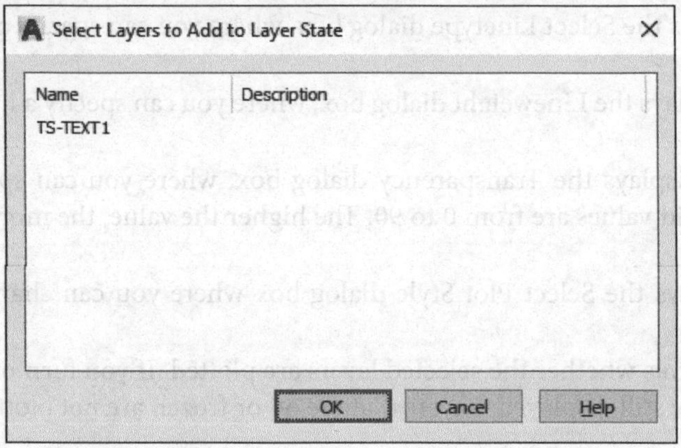

Fig. 4.36: Select Layers to Add to Layer State

Displays layers by name and includes descriptions that are not defined in the selected named layer state.

Multiple layers can be selected. Clicking OK displays, the selected layers in the Edit Layer State dialog box.

4.6.3.4 Edit Layer State Dialog Box

Modifies the selected **layer** state. **LAYER** (Command) Find: **Layer** States Manager > Edit Displays all the **layers** and their properties that are saved in the selected **layer** state.

Modifies the selected layer state. Displays all the layers and their properties that are saved in the selected layer state.

Layer List

Displays a list of layers in the selected layer state and their properties.

Delete

Removes the selected layers from the layer state.

Add

Displays the Select Layers to Add to Layer States dialog box, in which you can add layers from the current drawing that are not defined in the layer state.

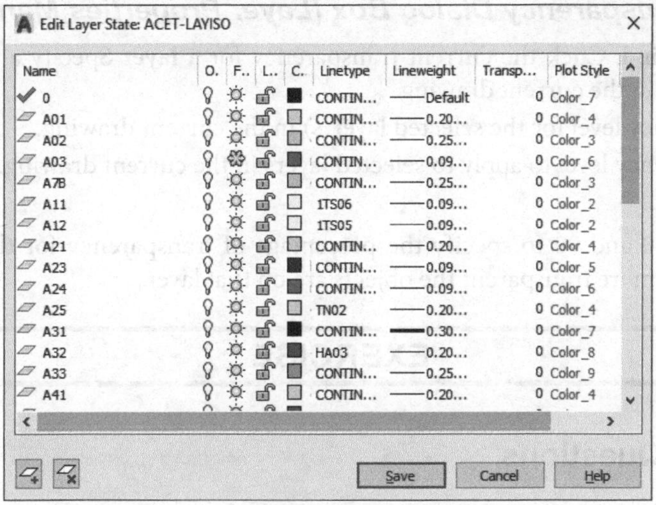

Fig. 4.37: Edit Layer State

4.6.3.5 Customize Layer Columns Dialog Box

Controls the display and order of columns in the **Layer** Properties Manager. **LAYER** (Command) Find: Right-click any **layer** column label > Customize List of options. The following options are displayed.

Controls the display and order of columns in the Layer Properties Manager.

Check the Columns you want to Display

Sets which columns are displayed in the Layer Properties Manager for the current Model tab or layout tab.

Move Up: Moves the selected column to the left in the Layer Properties Manager.

Move Down: Moves the selected column to the right in the Layer Properties Manager.

Select All: Selects all columns in the list.

Clear All: Clears selection from all columns in the list.

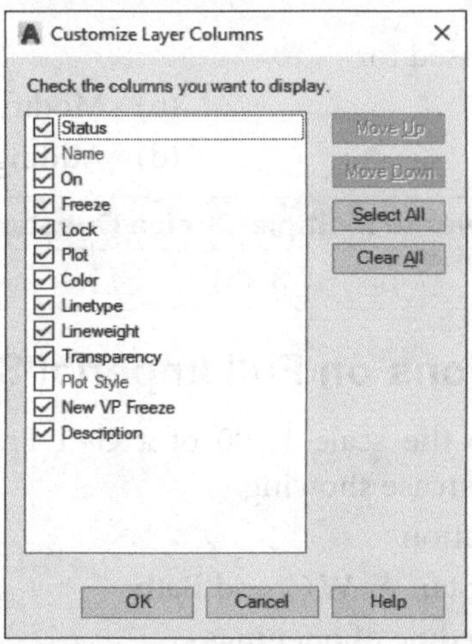

Fig. 4.38: Customize Layer Column

4.6.3.6 Layer Transparency Dialog Box (Layer Properties Manager)

LAYER (Command) Find: Click the current transparency for a **layer** Specify a transparency level to apply to selected **layers** in the current drawing.

Sets the transparency level for the selected layer(s) in the current drawing.

Specify a transparency level to apply to selected layers in the current drawing.

Transparency Value

Enter a value between 0 and 90 to specify the percentage of transparency for the selected layer. The higher the number, the more transparent the objects are on that layer.

EXERCISE

Multiple Choice Questions

1. Frame Structure is
   (a) Column Beam Structure
   (b) Open foundation
   (c) Brick foundation
   (d) Shallow foundation
2. Lintel Level is above
   (a) Terrace
   (b) Door Opening
   (c) At sill level
   (d) At D.P.C level
3. Sofit Level is
   (a) Above Door opening
   (b) In staircase
   (c) Beam Bottom
   (d) Below Window
4. R.C.C is
   (a) Reinforced Cement Concrete
   (b) Raw Cement Concrete
   (c) Red Cement Cone
   (d) Reinforced curb Concrete
5. Modify command in CAD is used for
   (a) Drawing
   (b) Modifying
   (c) Erasing
   (d) Adding

Answer to Multiple-Choice Questions				
1. (a)	2. (b)	3. (b)	4. (a)	5. (b)

Draw the Following Questions on Full Imperial Size Sheet (A1)

1. Draw submission drawing to the scale 1:100 of a G+1 Frame Structure residential building (2BHK) with flat Roof and staircase showing
   (a) Developed plan and elevation
   (b) Section passing through Stair or W.C. and Bath
   (c) Foundation plan and schedule of openings
   (d) Site plan (1:200), area statement, construction notes.

2. **Draw the above-mentioned drawing as Serial no.(B-1) using AutoCAD software and enclose the print out.**
   (a) Developed Plan
   (b) Elevation
   (c) Section passing through Stair or W.C. and Bath
   (d) c. Foundation plan and schedule of openings
   (e) d. Site plan (1:200), area statement, construction notes.
3. **Draw working drawing for above mentioned drawing as serial no (B-2) on AutoCAD showing?**
   (a) Foundation Plan to the scale 1:50
   (b) Detail enlarged section of R.C.C Column and footing with plinth filling
   (c) Detail enlarged section of R.C.C Beam, Lintel and Chajjas.
   (d) Detailed Enlarged Section of R.C.C Staircase and Slab.
4. **Draw on AutoCAD**
   (a) Section of the wall 300 mm thick with door opening
   (b) Section of the wall with window opening.

5 Perspective Drawing

UNIT SPECIFICS

Through this unit we have discussed the following aspects:

To Draw the Plan, Elevation, Section and Details of

- What is a perspective
- Types of Perspective
- Principles used in Perspective
- Examples of Two-point Perspective
- To explain drawing in 3 Dimension sketch

Scan Me

UNIT OUTCOMES

List of outcomes of this unit is as follows:

U5-O1: What is a perspective
U5-O2: Types of Perspective
U5-O3: Principles used in Perspective
U5-O4: Examples of Two-point Perspective
U5-O5: To explain drawing in 3 Dimension sketch

Unit-5 Outcomes	EXPECTED MAPPING WITH COURSE OUTCOMES (1- Weak Correlation; 2- Medium Correlation; 3- Strong Correlation)				
	CO-1	CO-2	CO-3	CO-4	CO-5
U5-O1	-	-	-	-	3
U5-O2	-	-	-	-	-
U5-O3	-	-	-	-	2
U5-O4	-	-	-	-	3
U5-O5	-	-	-	-	3

5.1 DEFINITION

Perspective is a three-dimensional view of any object seen by an observer. It is a Latin word meaning look through, Greeks had developed sense of observation regarding vanishing of object to a certain point. Ancient Indian architects has realized and achieved the desired optical effect. In 15th century this art was taken into consideration. Filippo Brunelleschi (1377-1446) of Renaissance period mentioned about the mathematical laws of perspective. This art was used by artists and painters to create realistic view of paintings, illusion of depth was created by the shadows, shading and light. Later it was adopted by Architects, Engineers as it helps to visualize any object like box, house, pillar, table, chair etc. in 3 Dimension representation. With the help of perspective, Architects can convey the true image of a proposed structure. The method of drawing the perspective is simple however it should be understood well before use. The process of drawing three-dimensional view on a two-dimensional paper is an art.

5.2 TYPES OF PERSPECTIVE

1. One Point Perspective or Parallel Perspective
2. Two Point Perspective or Oblique or angular Perspective
3. Three Point Perspective

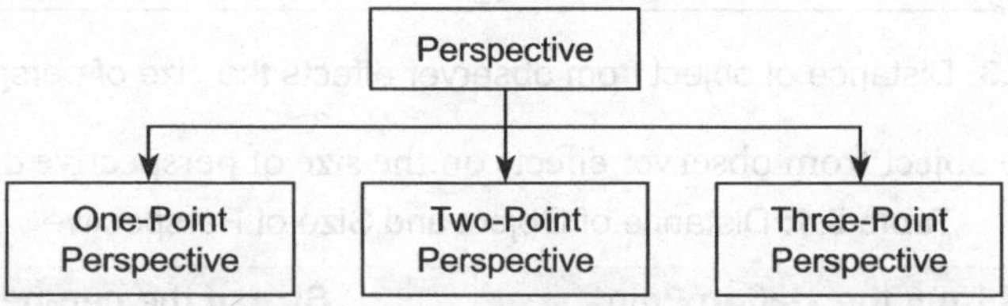

Fig. 5.1: Types of Perspective

Scan Me Scan Me Scan Me

5.3 TERMS USED IN MAKING PERSPECTIVE

5.3.1 Picture Plane

It is a two-dimensional plane on which three-dimensional building is drawn. It is an imaginary, transparent vertical plane erected between the object and the observer, very near to, or touching or cutting one corner of the object. To draw perspective of a small objects the plane can be behind the object.

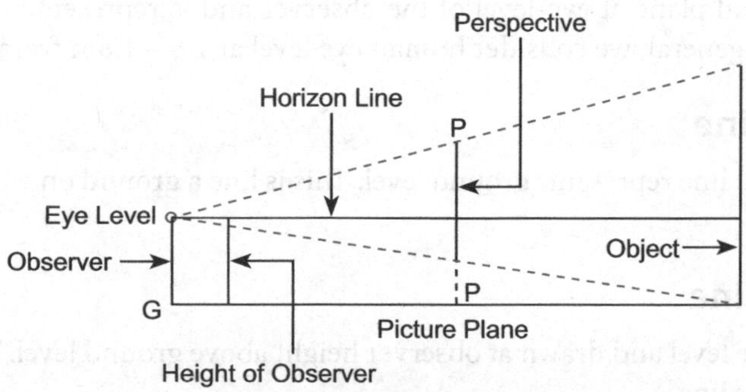

Fig. 5.2: Position of Picture Plane

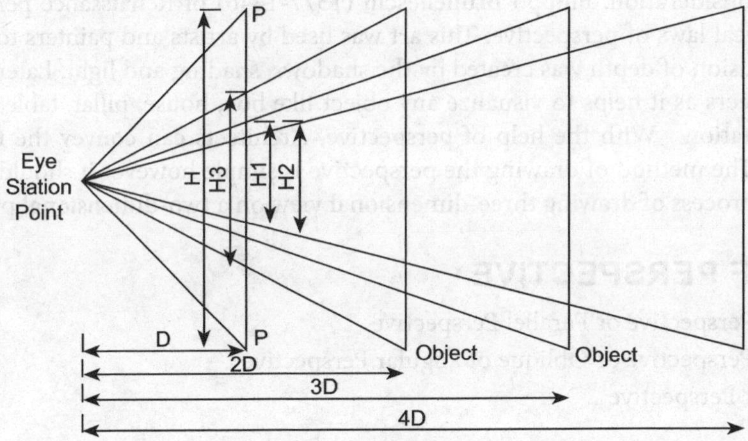

Fig. 5.3: Distance of object from observer affects the size of perspective

The distance of the object from observer effects on the size of perspective drawings.

Table 5.1: Distance of Object and Size of Perspective

Distance of object from the station Point	Size of the perspective of object
2D	H/2
3D	H/3
4D	H/4
ND	H/N

5.3.2 Eye Point or Station Point

It is the position of the observer in the Plan. It is station point **S** or Spectator Point in the Plan, whereas in elevation it is eye level. It is a point from which object is viewed. It is always in the front of object as well as picture plane. It can be in the center of the object or towards right or left of the object. It should be chosen in such a way that it should give a real pictorial view. It is common to fix the station point at 3.5 times the height of the object from the nearest corner or the object. Too close station point will give distorted view and too far will give very small perspective view.

5.3.3 Eye Level

It is a horizontal line and plane at eye-level of the observer and is represented by a straight line called Horizon Line (HL). In general, we consider human eye level at 1.6 – 1.8m from ground level.

5.3.4 Ground Line

It is a straight horizontal line represents ground level. This is line a ground on which observer is standing and visualizing.

5.3.5 Horizon Line

It is a line represents eye level and drawn at observer height above ground level. The objects convergence occurs at a point on this line.

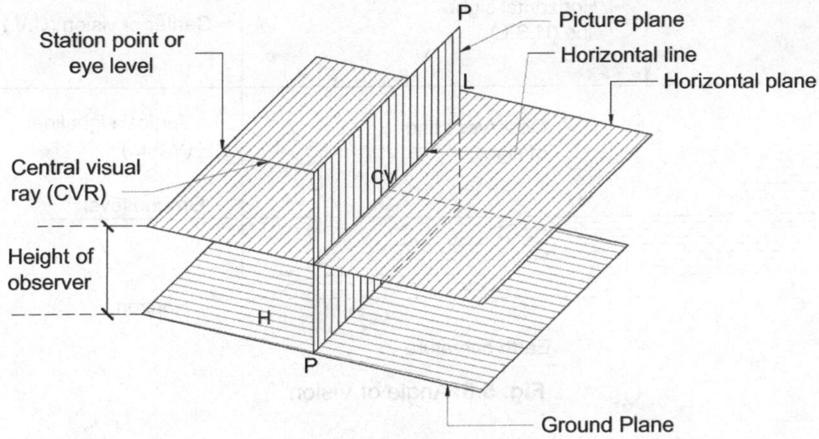

Fig. 5.4: Different Planes

5.3.6 Horizontal Plane

Horizontal plane is a plane passes through eye level. A line in this plane through picture plane is called horizon line as shown **Fig. 5.4.**

5.3.7 Centre of Vision

Vertical sight lines and horizontal sight lines are always at right angles to each other and the point where they cross is the centre of vision (CV), the point towards infinity which is dead in line with the observer's eye as shown **Fig. 5.5** It is also called point of sight or principal vanishing point.

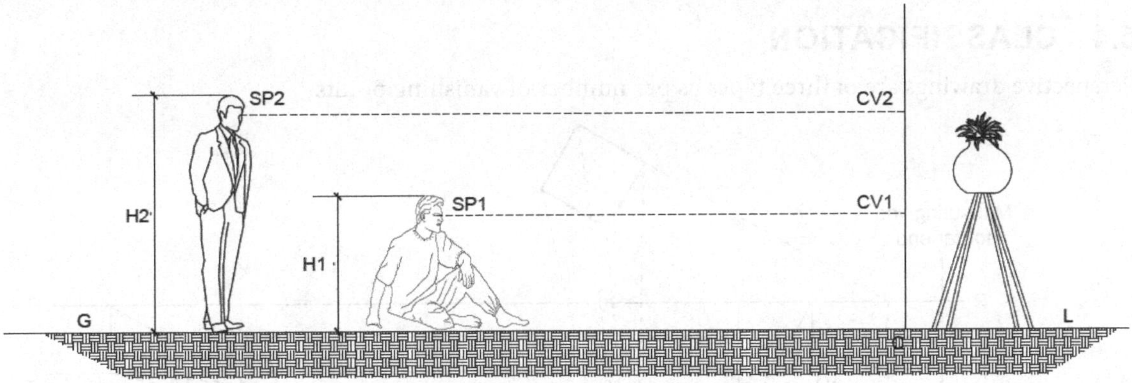

Fig. 5.5: Point of Sight

5.3.8 Angle of Vision

One can see clearly only a certain portion. Angle subtended by a horizontal plane and vertical direction in which someone can visualize the things clearly is called angle of vision. Horizontally angle of vision is generally 60 degrees and vertical angle of vision is 45degree. If the object lies in this area observer can visualize it, otherwise it is difficult to observe or cannot observe.

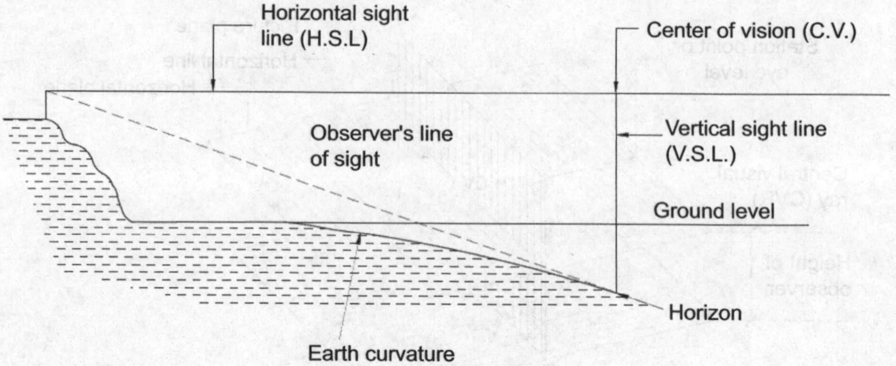

Fig. 5.6: Angle of Vision

5.3.9 Vanishing Point and Vanishing Lines

The main feature of the perspective drawing is vanishing of all parallel lines of object to a point to an infinite distance. The apparent convergence of the parallel lines is called vanish. The imaginary points where all rays vanish are called vanishing point and all imaginary lines which appear to be converging are called vanishing lines. These vanishing points lie at horizon plane at eye level.

5.3.10 Measure Line

The intersection of the plan of object with the picture plane gives true height at point of intersection when measured in elevation. This line is called measuring or measure line. The part of object touching the picture plane, their perspective height is always less than true height as shown in **Fig 5.7.**

5.4 CLASSIFICATION

Perspective drawings are of three types as per number of vanishing points

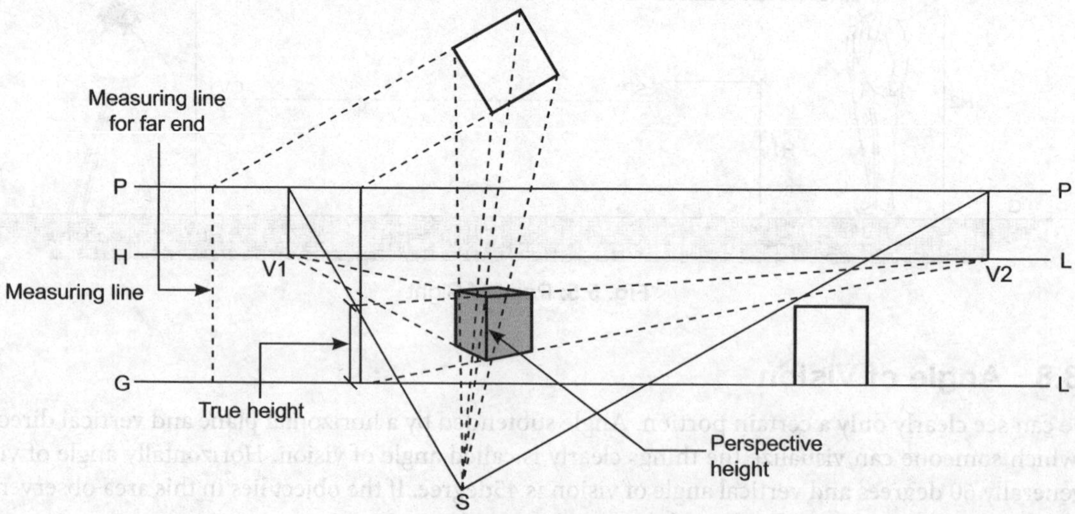

Fig. 5.7: Terms used in Perspective

1. One Point Perspective
2. Two Point Perspective
3. Three Point Perspective

In One Point perspective there is only one vanishing point. The perspective is also called as parallel perspective. In this type of perspective two sets of lines i.e., horizontal, and vertical lines are parallel to picture plane. That is why front section of the object shows true height and width and depth lines of the object converge towards one vanishing points as shown in the **Fig. 5.8**. Perspective of vertical and horizontal lines parallel to picture plane remain vertical and horizontal whereas horizontal lines perpendicular to picture plane convergence to center of vision i.e., Principal vanishing point. This type of perspective is drawn to show street vistas or interior view of the room.

Two Point Perspective: In two-point perspective the object is inclined to the picture plane. That is why only vertical lines remain parallel to picture plane whereas other sets of two lines are inclined to picture plane. Hence there are two vanishing points one for each set of horizontal lines. All lines going towards left of the plan converge towards left vanishing point and all lines going towards right on the plan converge towards right vanishing point. Vertical line remains vertical as shown in **Fig. 5.9.** Two-point perspective is used to show low height buildings and surroundings, shape of engineering parts. In this perspective one can see two faces of building in perspective drawing. This is a very common drawn perspective in showing the buildings. In case of multi storied buildings, even vertical lines which are parallel to picture plane also converge at the point just above the building forming third vanishing point. Even for small objects if the object is visualized in inclined position vertical lines seems to converge at a certain point. So, in **three Point** perspective there are three vanishing points and all lines of one set converge to one vanishing point as shown in **Fig. 5.10**. This is used to make tall buildings or multi storied view because two-point perspective for tall buildings give a distorted perspective view.

Following are the rules of the geometry of the perspective

1. The Perspective of the parallel lines inclined to the picture plane converge to a vanishing point.
2. The Perspective of the horizontal parallel lines inclined to the picture plane converge to a vanishing point lying on the horizon line (HL).
3. The Perspective of the lines perpendicular to the picture plane converge to a point the center of vision.
4. Perspective of vertical lines are vertical.
5. The Perspective of the horizontal lines which are parallel to picture plane are horizontal.

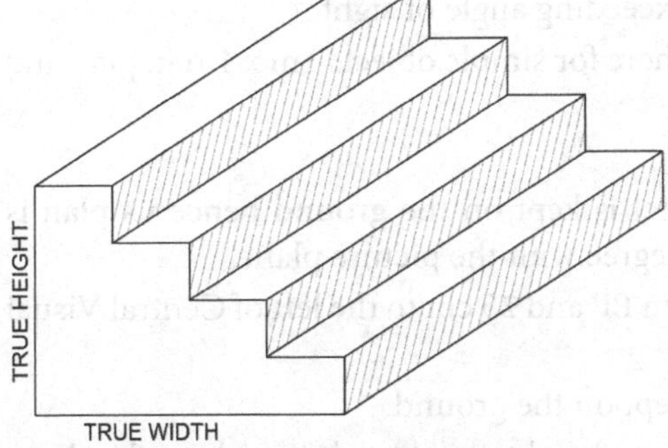

Fig. 5.8: One Point Perspective

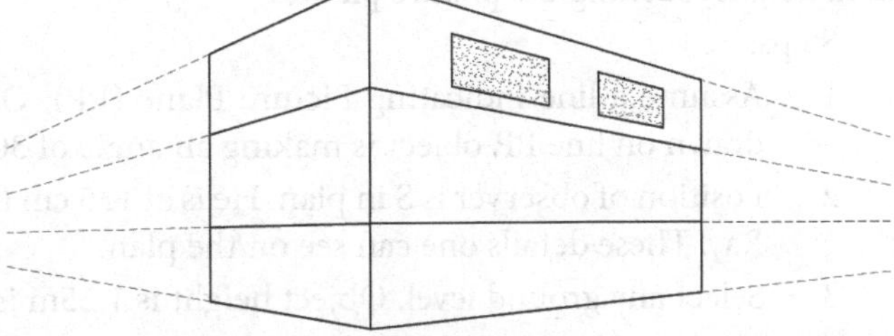

Fig. 5.9: Two Point Perspective

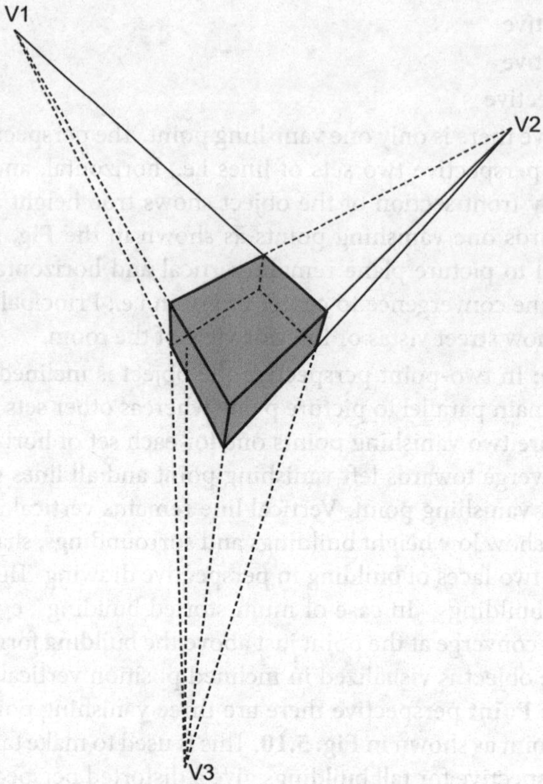

Fig. 5.10: Three Point Perspective

5.5 PERSPECTIVE DRAWINGS

Two Point perspective is very to make buildings. In drawing any perspective, the selection of different parameters should be appropriate, otherwise it will give distorted view. The common causes of distortion are:

- The view of the perspective from a wrong station point
- Incorrect projection of the object upon the picture plane
- The projection of an uncommon station point by exceeding angle of sight.

Method of drawing **Two Point** perspective is explained here for simple objects 1m x 1m in plan and 1.2 m height touching the picture plane.

Steps:

1. Assume a line indicating Picture Plane (PP). Object is kept on the ground hence its plan is drawn on line PP. object is making an angle of 30 degree with the picture plane.
2. Position of observer is S in plan. He is at 135 cm from PP and 25 cm to the left of Central Visual Ray. These details one can see on the plan.
3. Select any ground level. Object height is 1.25m is kept on the ground.
4. Draw one horizontal line (HL) above ground level at a height equal to observer's eye level i.e., 1.5m.

5. From station point S draw one line parallel to AD and other line parallel to AB. These lines cut PP at W1 and W2. Draw vertical lines from W1 and W2 downwards which cuts PP at V1 and V2, these are the vanishing points where objects horizontal and vertical lines converge.
6. From S join all the points of object up to PP and then project them vertically downward up to ground level (GL).
7. Point A touches the PP hence true height and perspective height of this point will be same. Transfer true height of the object of this line and converge them to vanishing points V1 and V2. These converging lines cut vertical projections of B and D at B1 B2 and D1D2 (upper and lower points of object). Hence sides A1 D1 A2 D2 and A1 B1 and A2 B2 are drawn **Fig 5.11**.
8. Similarly converge D1 D2 and B1 B2 to get C1 C2

True height of point B and D can be measured on measuring lines drawn for B and D separately.

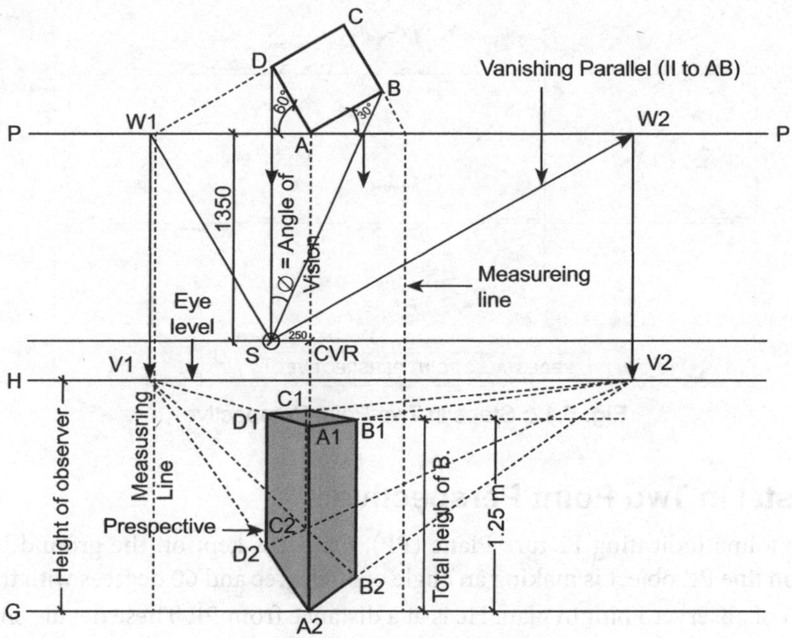

Fig. 5.11: Two Point Perspective

5.5.1 Some examples of Two Point Perspective

1. Assume a line indicating Picture Plane (PP). Object is kept on the ground hence its plan is drawn on line PP. object is making an angle of 45 degree with the picture plane.
2. Position of observer is Spectator Point in plan. He is at 175 cm from PP. These details one can see on the plan.
3. Select any ground level. Object height is 60cm is kept on the ground.
4. Draw one horizontal line (HL) above ground level at a height equal to observer's eye level i.e. 1.5m.
5. From station point S draw one line parallel to AD and other line parallel to AB. These lines cut PP at W1 and W2. Draw vertical lines from W1 and W2 downwards which cuts PP at V1 and V2, these are the vanishing points where objects horizontal and vertical lines converge.

6. from S join all the points of object up to PP and then project them vertically downward up to ground level (GL).
7. point A touches the PP hence true height and perspective height of this point will be same. Transfer true height of the object of this line and converge them to vanishing points V1 and V2. These converging lines cut vertical projections of B and D at B1 B2 and D1D2 (upper and lower points of object). Hence sides A1 D1 A2 D2 and A1 B1 and A2 B2 are drawn **Fig 5.12.**
8. similarly converge D1 D2 and B1 B2 to get C1 C2

True height of point B and D can be measured on measuring lines drawn for B and D separately.

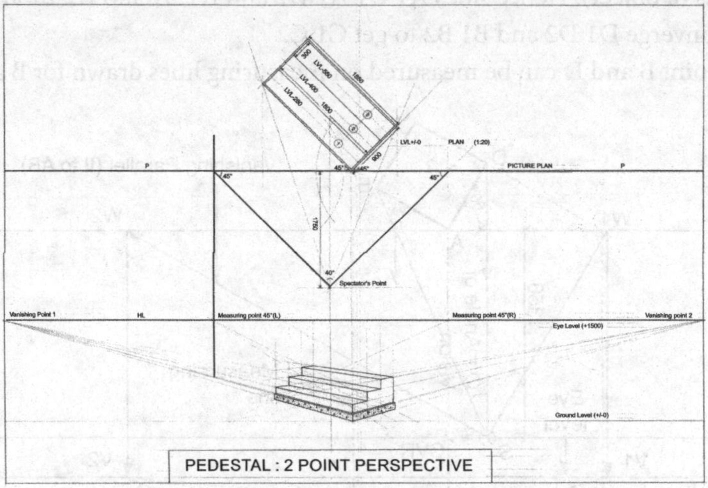

Fig. 5.12: Steps in Two Point Perspective

5.5.2 Pedestal in Two Point Perspective

1. Assume a line indicating Picture Plane (PP). Object is kept on the ground hence its plan is drawn on line PP. object is making an angle of 30 degree and 60 degrees with the picture plane.
2. Position of observer Point in plan. He is at a distance from PP. These details one can see on the plan.
3. Select any ground level. Object is kept on the ground.
4. Draw one horizontal line (HL) above ground level at a height equal to observer's eye level i.e 1.8m.
5. From station or observer point draw one line parallel to AD and other line parallel to AB. These lines cut PP at W1 and W2. Draw vertical lines from W1 and W2 downwards which cuts PP at V1 and V2, these are the vanishing points where objects horizontal and vertical lines converge.
6. From S join all the points of object up to PP and then project them vertically downward up to ground level (GL).
7. Point A touches the PP hence true height and perspective height of this point will be same. Transfer true height of the object of this line and converge them to vanishing points V1 and V2. These converging lines cut vertical projections of B and D at B1 B2 and D1D2 (upper and lower points of object). Hence sides A1 D1 A2 D2 and A1 B1 and A2 B2 are drawn **Fig 5.13.**

8. Similarly converge D1 D2 and B1 B2 to get C1 C2
True height of point B and D can be measured on measuring lines drawn for B and D separately.

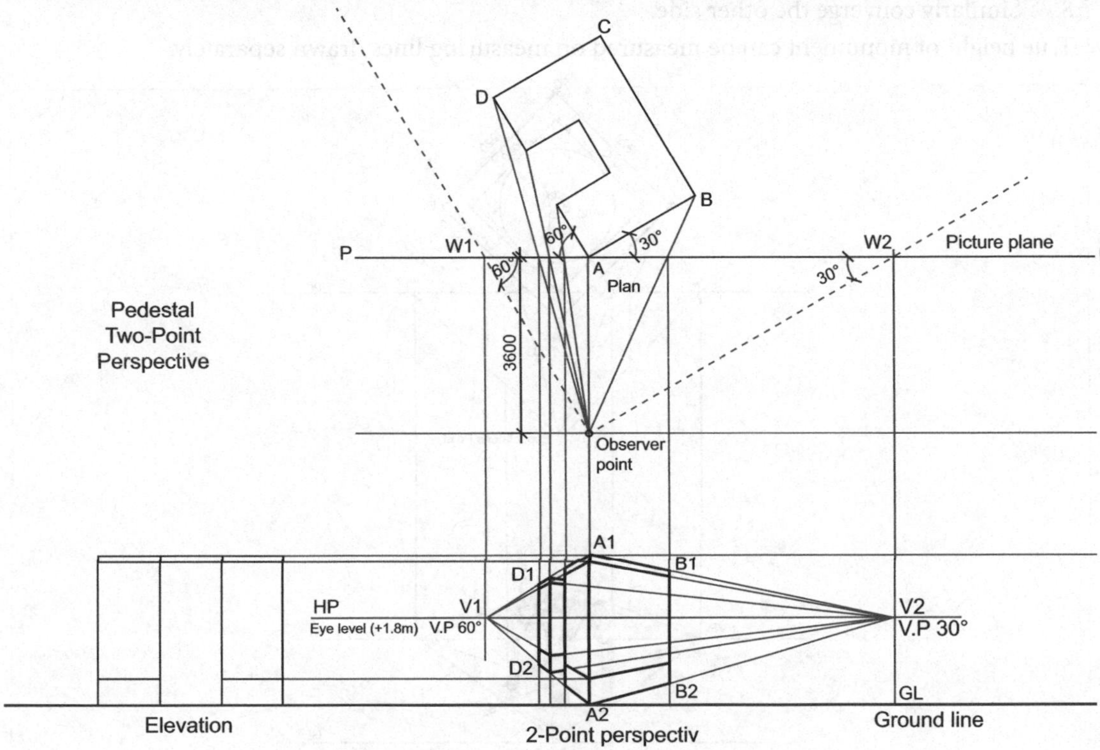

Fig. 5.13: Pedestal in Two Point Perspective

5.5.3 Monument in Two Point Perspective

1. Assume a line indicating Picture Plane (PP). Object is kept on the ground hence its plan is drawn on line PP. object is making an angle of 30 degree and 60 degrees with the picture plane.
2. position of observer Point in plan. He is at a distance from PP. These details one can see on the plan.
3. Select any ground level. Object is kept on the ground.
4. Draw one horizontal line/ Elevational Line (HL or EL) above ground level at a height equal to observer's eye level i.e. 1.8m.
5. From station or Spectator point draw one line parallel to AD and other line parallel to AB. These lines cut PP at W1 and W2. Draw vertical lines from W1 and W2 downwards which cuts PP at V.P.1 and V.P.2, these are the vanishing points where objects horizontal and vertical lines converge.
6. From S join all the points of object up to PP and then project them vertically downward up to ground level (GL).
7. Point A touches the PP hence true height and perspective height of this point will be same. Transfer true height of the object of this line and converge them to vanishing points V1 and

V2. These converging lines cut vertical projections (upper and lower points of object). Hence all the other sides are drawn **Fig 5.14.**

8. Similarly converge the other side.

True height of monument can be measured on measuring lines drawn separately.

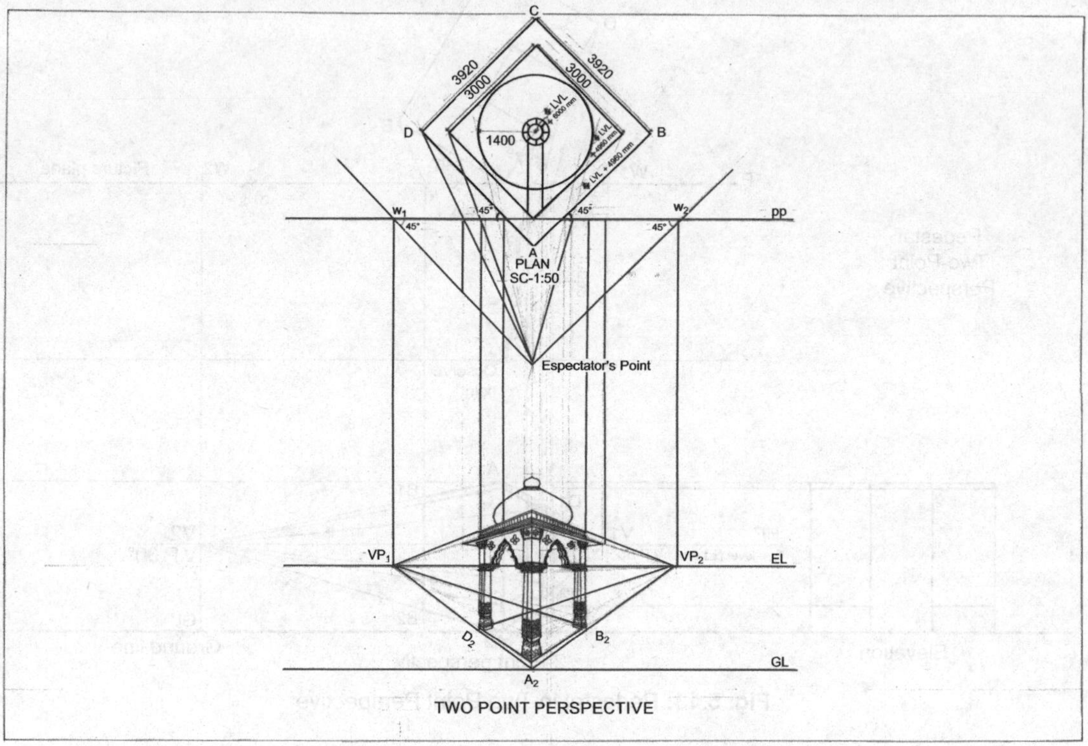

Fig. 5.14: Monument in Two Point Perspective

EXERCISE

Multiple Choice Questions

1. Perspective is
   (a) Three-Dimensional View of the object.
   (b) Two-Dimensional View of the object.
   (c) Top View of the object.
   (d) Side View of the object.
2. For bungalow Architects make
   (a) One Point Perspective
   (b) Two Point Perspective
   (c) Three Point Perspective
   (d) None

3. Eye Level Is at
   - (a) 2.3 – 2.8m
   - (b) 1.3-1.8m
   - (c) 0.9-1.2m
   - (d) 2.5-2.9m
4. Station Point
   - (a) Observer Point
   - (b) Eye Point
   - (c) Spectator Point
   - (d) All the Above
5. Ground Line is
   - (a) Observer Standing
   - (b) Eye level
   - (c) Vanishing Line
   - (d) Picture plane

Answer to Multiple-Choice Questions				
1. (a)	2. (b)	3. (b)	4. (d)	5. (a)

Answer the Following Questions

1. What is Perspective?
2. How many types of Perspective view are there?
3. Draw two-point perspective drawing of small objects - monuments, scale 1:50
   - (a) Draw plan, elevation, eye level, picture plane and vanishing points
   - (b) Draw perspective view.
4. Draw two-point perspective drawing of small objects - steps, scale 1:50
   - (a) Draw plan, elevation, eye level, picture plane and vanishing points
   - (b) Draw perspective view.
5. Draw two-point perspective drawing of small objects - pedestals scale 1:50
   - (a) Draw plan, elevation, eye level, picture plane and vanishing points
   - (b) Draw perspective view.

REFERENCES

1. Shah. M.G. Kale, CM, Patki, S.Y., Building Drawing, McGraw Hill Publishing company Ltd. New Delhi.
2. Walters V Nigel, Bromham John, Principles of Perspective, The University Press, Glasgow.
3. Mulik Shankar, Perspective and Sciography, Allied Publisher Limited. Bombay.
4. Deodhar S.V, Building Drawing, New Vrinda Publishing House Jalgaon.

CO AND PO ATTAINMENT TABLE

Course outcomes (COs) for this course can be mapped with the programme outcomes (POs) after the completion of the course and a correlation can be made for the attainment of POs to analyze the gap. After proper analysis of the gap in the attainment of POs necessary measures can be taken to overcome the gaps.

Table for CO and PO attainment

Course Outcomes	ATTAINMENT OF PROGRAMME OUTCOMES (1- Weak Correlation; 2- Medium Correlation; 3- Strong Correlation)						
	PO-1	PO-2	PO-3	PO-4	PO-5	PO-6	PO-7
CO-1							
CO-2							
CO-3							
CO-4							
CO-5							

The data filled in the above table can be used for gap analysis.

INDEX